SCIENTISTS OF WALES

Joseph
Harris

SCIENTISTS OF WALES

Series Editors
Gareth Ffowc Roberts
Bangor University

John V. Tucker
Swansea University

Iwan Rhys Morus
Aberystwyth University

Previously published in the series

Gordon Roberts, *Robert Recorde: Tudor Scholar and Mathematician* (2016)

Iwan Morus, *William Robert Grove: Victorian Gentleman of Science* (2017)

Rowland Wynne, *Evan James Williams: Ffisegydd yr Atom* (2017)

John Baylis, *Wales and the Bomb: The Role of Welsh Scientists and Engineers in the British Nuclear Programme* (2019)

Rowland Wynne, *Evan James Williams: Atomic Physicist* (2020)

Nicola Bruton Bennetts, *William Morgan: Eighteen Century Mathematician, Scientist and Radical* (2020)

Brynley F. Roberts, *Edward Lhwyd: c.1660–1709, Naturalist, Antiquary, Philologist* (2022)

Haydn Edwards, *Griffith Davies: Arloeswr a Chymwynaswr* (2023)

Gavin Gatehouse, *Griffith Evans, 1835–1935: Veterinarian, Pioneer Parasitologist and Adventurer* (2024)

Paul Frame, *'A World of New Ideas': 1650–1820, Welsh Scientists of the Long Eighteenth Century, Volume 1: The Isles* (2025)

SCIENTISTS OF WALES

Joseph Harris

SCIENTIST, ARTISAN, ASSAY MASTER

MARTIN GRIFFITHS

UNIVERSITY OF WALES PRESS
2025

© Martin Griffiths, 2025

All rights reserved. No part of this book may be reproduced in any material form (including photocopying or storing it in any medium by electronic means and whether or not transiently or incidentally to some other use of this publication) without the written permission of the copyright owner except in accordance with the provisions of the Copyright, Designs and Patents Act 1988. Applications for the copyright owner's written permission to reproduce any part of this publication should be addressed to the University of Wales Press, University Registry, King Edward VII Avenue, Cardiff CF10 3NS.

www.uwp.co.uk

British Library Cataloguing-in-Publication Data
A catalogue record for this book is available from the British Library.

ISBN   978-1-83772-300-3
eISBN   978-1-83772-301-0

The right of Martin Griffiths to be identified as author of this work has been asserted in accordance with sections 77, 78 and 79 of the Copyright, Designs and Patents Act 1988.

For GPSR enquiries please contact: Easy Access System Europe Oü, 16879218 Mustamäe tee 50, 10621, Tallinn, Estonia. *gpsr.requests@easproject.com*

The University of Wales Press gratefully acknowledges the support of the Books Council of Wales in publishing this title.

Typeset by Marie Doherty
Printed and bound by CPI Group (UK) Ltd, Croydon, CR0 4YY

*This book is dedicated to my wonderful
grandchildren Morwenna and Macsen*

*May they be proud of their Welsh heritage and its
hidden figures that changed the world*

CONTENTS

Series Editors' Preface	ix
Acknowledgements	xi
List of Illustrations	xiii
Introduction	1
1 Joseph's Early Life	3
2 London and Apprenticeship	15
3 The Caribbean Voyages	31
4 Navigation, Cartography and Mathematics	47
5 Assay Master at the Mint	61
6 Joseph's Family and Relationships	73
7 The Carysfort Commission	95
8 The Transit of Venus	103
9 Joseph's Economic Influence and Essay on Money and Coins	119
10 Joseph's Final Scientific Treatise	127
11 Joseph's Legacy and His Heirs	135
Appendix A: Genealogy of Joseph Harris	147
Appendix B: Timeline of Joseph Harris	149
Notes	151
Bibliography	157
Index	161

SERIES EDITORS' PREFACE

Wales has a long and important history of contributions to scientific and technological discovery and innovation stretching from the Middle Ages to the present day. From medieval scholars to contemporary scientists and engineers, Welsh individuals have been at the forefront of efforts to understand and control the world around us. For much of Welsh history, science has played a key role in Welsh culture: bards drew on scientific ideas in their poetry; renaissance gentlemen devoted themselves to natural history; the leaders of early Welsh Methodism filled their hymns with scientific references. During the nineteenth century, scientific societies flourished and Wales was transformed by engineering and technology. In the twentieth century the work of Welsh scientists continued to influence developments in their fields.

Much of this exciting and vibrant Welsh scientific history has now disappeared from historical memory. The aim of the Scientists of Wales series is to resurrect the role of science and technology in Welsh history. Its volumes trace the careers and achievements of Welsh investigators, setting their work within their cultural contexts. They demonstrate how scientists and engineers have contributed to the making of modern Wales as well as showing the ways in which Wales has played a crucial role in the emergence of modern science and engineering.

RHAGYMADRODD GOLYGYDDION Y GYFRES

O'r Oesoedd Canol hyd heddiw, mae gan Gymru hanes hir a phwysig o gyfrannu at ddarganfyddiadau a menter gwyddonol a thechnolegol. O'r ysgolheigion cynharaf i wyddonwyr a pheirianwyr cyfoes, mae Cymry wedi bod yn flaenllaw yn yr ymdrech i ddeall a rheoli'r byd o'n cwmpas. Mae gwyddoniaeth wedi chwarae rôl allweddol o fewn diwylliant Cymreig am ran helaeth o hanes Cymru: arferai'r beirdd llys dynnu ar syniadau gwyddonol yn eu barddoniaeth; roedd gan wŷr y Dadeni ddiddordeb brwd yn y gwyddorau naturiol; ac roedd emynau arweinwyr cynnar Methodistiaeth Gymreig yn llawn cyfeiriadau gwyddonol. Blodeuodd cymdeithasau gwyddonol yn ystod y bedwaredd ganrif ar bymtheg, a thrawsffurfiwyd Cymru gan beirianneg a thechnoleg. Ac, yn ogystal, bu gwyddonwyr Cymreig yn ddylanwadol mewn sawl maes gwyddonol a thechnolegol yn yr ugeinfed ganrif.

Mae llawer o'r hanes gwyddonol Cymreig cyffrous yma wedi hen ddiflannu. Amcan cyfres Gwyddonwyr Cymru yw i danlinellu cyfraniad gwyddoniaeth a thechnoleg yn hanes Cymru, â'i chyfrolau'n olrhain gyrfaoedd a champau gwyddonwyr Cymreig gan osod eu gwaith yn ei gyd-destun diwylliannol. Trwy ddangos sut y cyfrannodd gwyddonwyr a pheirianwyr at greu'r Gymru fodern, dadlennir hefyd sut y mae Cymru wedi chwarae rhan hanfodol yn natblygiad gwyddoniaeth a pheirianneg fodern.

ACKNOWLEDGEMENTS

This book is the culmination of almost two decades of getting to know Joseph Harris and his influence on our history and economy. I am indebted to the Trefecca College for access to Joseph's telescopes and his letter on the Transit of Venus. I am also incredibly indebted to Jenny Stanesby-Moody for her help in accessing the telescope, in giving me permission to include her translation of Joseph's letter to the Royal Society, and for allowing me access to her beautiful home and her patient friendship and encouragement in this endeavour.

I am also grateful for the help of the National Library of Wales for allowing me access to the Harris family letters and the diaries of Howell Harris.

Martin Griffiths

LIST OF ILLUSTRATIONS

Figure 1 St Gwendoline's church at Talgarth (photograph by author) 4

Large stone-built church with a porch on the left side and a tall tower at the left-hand end. The church is red in colour as it is built using the local old red sandstone. Inside the church are the memorial tablets to Joseph and Howell Harris.

Figure 2 Llwyn Llwyd House, Joseph and Howell's school (photograph by author) 9

Large stone house, a former barn with an extension on the left-hand side of the door, a slate roof and wooden windows. The house stands in a large courtyard.

Figure 3 Tredustan Court (photograph by author) 12

Large eighteenth-century manor house of two storeys in an L-shape with the entrance at right-hand side. Built in local red sandstone with mullioned windows and extensive garden containing a well.

Figure 4 Library at Tredustan Court (photograph by author) 12

Large library with a baby grand piano in the foreground covered in a shawl flanked by small chairs with two reading lamps in the background. Behind the piano in the background are floor-to-ceiling bookshelves with volumes of books filling all shelves. A fireplace is at the left-hand side with an armchair close by and a table.

Figure 5	Solar eclipse broadsheet by Senex (courtesy of the Royal Society)	13
	Large single piece of paper with a map of the British Isles and the pathway of the 1715 solar eclipse across it from north-east to south-west and the area of greatest eclipse drawn in a darker hue.	
Figure 6	Orrery at former Royal Society Meeting place Crane Court (photograph by author)	20
	A small model of a solar system hanging over an arched entrance to Crane Court. The model is simply a globe with smaller globes fixed in place surrounding it.	
Figure 7	Halley's eight-foot quadrant (photograph by author)	21
	Large wall-mounted quadrant (a quarter of a circle divided into 90 degrees) constructed of an iron lattice with a graduated quarter circle attached. A quadrant is in place at the Royal Greenwich Observatory with an attached sighting tube which can be moved up and down the quadrant to measure the altitude of stars above the horizon.	
Figure 8	Portrait of Edmond Halley (courtesy of the Royal Society)	26
	The Astronomer Edmond Halley in profile wearing a white wig and brown jacket. Underneath is a white shirt with a white cravat signifying that he has the religious title of Reverend, which was common to University graduates of his day. Halley is about 40 years old. Painted for the National Portrait Gallery.	
Figure 9	Gosfield Hall (photograph by author)	44
	Imposing white façade of the front of Gosfield Hall with Georgian windows and archways in Palladian style. The roof is penetrated by Mansard windows.	
Figure 10	Harris's southern sky map (courtesy of the Royal Greenwich Observatory)	49
	Joseph Harris's map of the southern sky with constellations drawn on it in classical design of mythological creatures and	

stars showing different brightnesses. Joseph's name appears at the top right-hand corner of the map above the date.

Figure 11 Devereux Tower, Tower of London (photograph by author) 69

Devereux Tower stands at the corner of the tower of London to the left of the visitor entrance. It is a twin tower design from the sixteenth century in limestone, separate from the white tower and overlooking the buildings where the Royal Mint formerly stood.

Figure 12 St Benet's church outside sign (photograph by author) 75

The sign is in gold lettering on a blue background. Close to St Paul's cathedral, the church itself is set right on the bank of the river Thames and is one of the churches designed by Sir Christopher Wren after the 1666 great fire of London. It is designed in red brick and white cornerstones with a tall tower surmounted by a dome and lantern coated in lead.

Figure 13 Howell Harris (courtesy and © The History Collection/Alamy) 77

Figure 14 Trefecca College today (photograph by author) 78

Front of Trefecca college with a motto surmounting the door and the front having gabled projections. The college is painted white and has a central red doorway.

Figure 15 Howell's musket and swords (photograph by author) 89

A glass case containing two long musket rifles and two swords. One is a ceremonial officer's sword with scabbard and the other is a rapier without a scabbard.

Figure 16 Howell's memorial tablet (photograph by author) 90

The memorial tablet of Howell Harris. It is a rectangular slate with inlaid gold lettering recounting his ministry and a short description of his life.

Figure 17	Transit of Venus (photograph by author)	104
	The sun in yellow with a black dot traversing it. The black dot is the planet Venus. Photographed by the author during the 2004 Transit of Venus.	
Figure 18	Joseph's telescope under investigation (photograph by author)	108
	Joseph's five-foot focal-length telescope laid on a trestle table with protective tissue underneath the tube. The author is on the far left, Jenny Stanesby-Moody is at centre, with Professor Peter Duffet-Brown on the right. Flanking Jenny are Richard and Mair, the secretaries of Trefecca College.	
Figure 19	Total lunar eclipse (photograph by author)	110
	The moon in total eclipse. The moon is full but is stained red by the passage of sunlight through the atmosphere.	
Figure 20	Joseph's telescope in a glass case at Trefecca (photograph by author)	111
	Joseph's five-foot telescope, one of the instruments used to observe the 1761 transit of Venus. The telescope is in a glass case and is mounted high on a wall of the Trefecca Museum.	
Figure 21	Joseph's letter to the RS (photograph by author)	113
	Large board with the pages of Joseph's letter to the Royal Society with his observations of the eclipse laid out at Trefecca College.	
Figure 22	Joseph's memorial tablet in St Gwendoline's church (photograph by author)	141
	On ornate memorial tablet in slate background with a white alabaster vase surmounting a shield with Joseph's achievements highlighted against the alabaster in black lettering. Above the shield and just below the vase is an image of a quadrant and a pair of mathematical compasses.	

INTRODUCTION

Wales may be a small nation, but it has a wonderful sense of history; a language and culture that is found nowhere else; a sense of pride in the past and a strong perception of its independence, all married to an incredible diversity of landscape. Although its population has remained relatively small, the contributions of its people to the wider world is out of proportion to their numbers and their impact on history, culture, commerce and sciences as found around us today. Much goes completely unnoticed, but can still be teased out by the interested reader.

Joseph Harris of Breconshire, the son of a carpenter and smallholder, gave of his time and energies in the pursuit of all these endeavours. His input has been rarely acknowledged or has been lost amongst the stories of great men who made substantial breakthroughs. His life story is a fine example of the cumulative nature of knowledge and the hidden people behind great discoveries. His contributions and standing were lamented by the historian Theophilus Jones in his *History of the County of Brecknock*:

> Of Joseph Harris, the eldest, who married one of the daughters, and heiress, of Thomas Jones, of Tredustan, little has been recorded beyond the information derived from his monument in the church. His talents were highly respectable, and indeed pre-eminent. But though he wrote several astronomical treatises, which are highly thought of, and was esteemed by the learned and great of his day, no biographer has written his life: no anecdotes of him have been preserved; nor have his virtues or talents been recorded farther

than as they appear in his works, which in general are anonymous. Indeed, that modesty, which is so amiable in him, seems to have descended to his posterity where he was born, for after all the enquiries I have made with respect of him, instead of learning any other particulars of his life, I have received only general encomiums and empty praise. I am much hurt that this self-taught philosopher, who was an honour to this county of Brecon, should pass almost unnoticed. The blame lies not with me, for it seems to have been destined, that his record should be only in heaven.

It seems pitiful to me that such a great man, with talents and attributes that were valued by the scientific and political establishment of his day, has almost nothing visible in this country to show his worth and legacy. There is a small memorial to him in Saint Gwendoline's church Talgarth, which says very little about his contributions to science, politics and the functioning of the economy right up to our modern day. His life can only be found in snippets of information gleaned from various sources. His interactions with his brothers through their letters both reveal and at the same time conceal his influence; it is known that he was a very modest man and would not push himself forward for recognition, not even placing his name in the first instance on the books that he would go on to write or the instruments he invented.

I hope therefore that this small volume will do much to illuminate the character and career of this wonderful, humble man, one of the truly remarkable if not great, sons of Wales.

1

JOSEPH'S EARLY LIFE

Joseph Harris was born to Howel and Susannah Harris (nee Powell) of 'Trefecca' (originally Trevecka), now Trefecca, a hamlet less than a mile from Talgarth in the former county of Breconshire, now part of Powys. Joseph's father, Howel ap Howell (also known as Powell) was a carpenter originally from Llangadog in Carmarthenshire and he moved to Trefecca as Susannah lived in the area. Parish records reveal their marriage on 5 September 1702. The couple went on to have a daughter, Anne and two more sons who would also become famous: Howell, for his nonconformist preaching and establishment of a religious community, eventually becoming known as the 'Apostle of Wales' by the Methodist community; and his younger brother Thomas, who enjoyed an illustrious career as a tailor and uniform maker to the British Army, eventually becoming the Sheriff of Breconshire. They also had another son, a youngster they also named Thomas, but little is known of this child and he may have died in infancy as none of the brothers mention him in their extensive letters later in life. It seems also that the daughter Anne did not survive into adulthood.

Why would these brothers become known as 'Harris'? It would appear that Howel ap Howell (Powell) changed his name shortly after the marriage to 'Harries' as it fitted better with his idea that the name would sound more anglicised. Church records at Talgarth, the small town close to Trefecca, show that the register records the change of name from Powell to Harries, eventually becoming known to us as Harris. Despite this apparent anglicisation, the family spoke Welsh at home though the brothers would correspond in English and no doubt

FIGURE 1 St Gwendoline's church at Talgarth (photograph by author)
Large stone-built church with a porch on the left side and
a tall tower at the left-hand end. The church is red in colour as
it is built using the local old red sandstone. Inside the church
are the memorial tablets to Joseph and Howell Harris.

their education and conversations at home would have been bilingual too.

There are no existing records giving further details of the day or month of Joseph's birth, but from church records it appears that his christening took place in Saint Gwendoline's church, Talgarth on 16 February 1703. It was the usual practice to baptise babies quickly as rates of infant mortality were very high in the early eighteenth century. It is my contention that Joseph was born probably in January 1704 by our modern reckoning. This reveals two things: that the family were probably Anglican in their general Christian belief and that Joseph was probably born in early 1704 by our modern calendar. This is due to the fact that there have been calendrical changes in our modern history. Before the introduction of the Gregorian calendar, the New Year in Britain started on 25 March and ran until 24 March the next year. Thus he would have been born in late 1703 by the eighteenth-century calendar but by our calendar this would be the year 1704. His grave tablet informs the reader that he was sixty-two years old when he died. If born in early 1704, this would put him in his sixtieth year at his death in September 1764.

However, historians Morgan Hugh Jones, John Edward Lloyd, Robert Thomas Jenkins and others state his birth as 1702. It has been postulated that the change over from the Julian to the Gregorian calendar is responsible for the confusion over the difference for Joseph's birth quoted by many authors. The Calendar Act of 1750 put the start of each year at the 1 January and subsequently we can assume that this would place Joseph's birth in early 1704 by modern reckoning and Joseph's death should be recorded at the age of sixty.[1]

Wales in the early eighteenth century

What was Wales like in the time of Joseph's upbringing? The country was relatively thinly populated; baptismal records suggesting a population around 400,000, fewer than 20 per cent of whom lived in towns of appreciable size. Most of Wales was a collection of hamlets and villages that passed for towns and remoteness and isolation remained a fact of life for most people, their horizons shaped by the immediate landscape around them. Although drovers' roads connected most places with markets at specific towns such as Carmarthen and even London, it was a long journey to get to the capital and roads did not greatly improve until the reign of George III at the latter part of the century. Most clothing, hardware and other necessities could be obtained at the local country fairs that were generally held in one location or another in the summer months, or at the local market towns.[2]

Industrially, Wales of the late seventeenth and eighteenth centuries was a non-contender on world markets or even on home ones. Despite the massive contributions of Wales to the industrial revolution in the latter half of the eighteenth century, in the earlier part of that century Wales had a latent but moribund industry in lead and copper mining, smelting and coal mining, leftovers from the Elizabethan age, though in general these were confined to the south Wales Valleys and the coast. Such industries were non-existent in rural mid Wales and certainly did not intrude upon the agricultural life of the Talgarth area, which was mostly dominated by the wool trade. Ironically, one of the growth industries of the time was tourism, many coming to experience the wild landscapes and gentle rolling topography of north and mid Wales. Often travellers would travel down the Wye river from Ross to Monmouth or Chepstow whilst others recorded the landscapes, outstanding buildings and the nature of the Welsh people in areas that were difficult to access from coach roads. Most of the written accounts of such travellers tended to speak negatively of the natives.[3]

One salient aspect of life in such a rural community was that local life and politics were dominated by the 'squires' who would have incomes and land that could total £500 or more. Many would be regarded as the local Justices of the Peace and would live in conditions that befitted their status. Some became the Sheriffs of their county and were looked upon as people who not only helped the local communities but also kept law and order and promulgated parliamentary statutes. In his published essay, *The Remaking of Wales in the 18th Century*, the historian Peter Thomas writes in 2010 that although there may be several hundred squires, at any one time, there were never more than twenty-seven Welsh MPs, so a small number of men and families controlled Wales and applied for parliamentary representation. Joseph of course was aware of this hierarchy and later in life he was selected by the local squire as a person worthy of attention.[4]

Although rural life may have stayed much as it was over the centuries, the same cannot be said for the state of science which was making rapid progress into understanding the natural world around us. Since the publication of Nicolaus Copernicus' *De Revolutionibus* in 1573 and the works of Johannes Kepler and his laws of planetary motion by 1630, the world was advancing to an ever-increasing sophistication of knowledge, facts and their application. Aided by various scientific instruments such as the telescope and microscope and more rational forms of enquiry that challenged the old ideas of astrology, alchemy and theology, science was becoming the pre-eminent way of enquiring into nature by the eighteenth century. In addition, the publication of *Philosophiae Naturalis Principia Mathematica* by Isaac Newton in 1687 put the physical world on a mathematical footing that could be checked, challenged and understood by its adherents.

Joseph was born into this rapidly changing world of wonder and understanding which we now call the 'Enlightenment', marked by the popularisation of science, the growth of scientific societies, the expansion of universities across Europe and a growing literate population. There was also a difference between knowledge as judged by the old order of things and claims to authority, such as those of the Catholic Church which through St Thomas Aquinas had welded church authority to Aristotelian doctrine that did not depend on empirical data but rather went by the ancient writings of such philosophers as Plato and Aristotle, whose view of the world had become largely accepted, unchallenged for centuries.

The Enlightenment introduced a new emphasis based on empirical knowledge, one grounded in experience. Such might include scientific experiments or observation and, for any proposition to be accepted as true, it must be verifiable and capable of practical demonstration. It had to be repeatable in front of witnesses. Enlightenment thinkers allowed theoretical or speculative thought as hypotheses that could guide experiments or primary observations, but they still demanded hard evidence to back such claims which, in their turn, would enter the pantheon of human knowledge as accepted fact.

Changes in our understanding of the natural world included: the discovery of oxygen and other elements that differed from the four traditional elements of Air, Earth, Fire and Water; the circulation of blood as understood by William Harvey and the heart as a pump rather than the seat of emotions; the separation of thinking processes from the heart to the brain; the production of vaccines by Edward Jenner; the invention of the steam engine and the transfer of manufacturing from cottage industries to manufactories all due to the power loom and the spinning jenny, inventions that transformed Britain into an industrial power.

Education and religion in rural Wales

Knowledge of the outside world was in many ways strictly limited to the greater population as newspapers and leaflets were confined to the border towns and written in English; such newspapers were read mostly by the gentry or professional classes. Most of the Welsh populace at the time would remain illiterate in English, though most could read and write Welsh. Welsh remained the lingua franca of most of Wales's regions and its religion either Anglican or staunchly nonconformist. During Joseph's lifetime, the Methodist revival would ensure the dominance of nonconformist religious ideology in Wales, and his brother Howell would play a large part in its establishment.

Books in Welsh were relatively scarce, though this was an improving situation under the later leadership of the Methodists, although most publications were religious tracts and booklets. The bible had been translated into Welsh as early as 1588 by Bishop William Morgan and the first book in Welsh, *Yny lhyvyr hwnn* by Sir John Price as early as 1564, but, due to price and general illiteracy, such books were not widely distributed.

Education in Wales was reserved for the hierarchy who generally went off to Oxford or Cambridge, a situation that did not improve for the less well off until the advent of schools known as dissenting academies, which were set up by the nonconformist Methodist religious movement to educate children across the UK and were amongst the first schools accessible to the general public. Organised and run by the dissenting movement that broke away from the Anglican church, these schools were the bedrock of the educational system in England and Wales from the mid seventeenth through to the late nineteenth century. Their style of teaching and the subjects taught were well developed by the time Joseph and his brothers attended in the early eighteenth century.

During Joseph's early life, changes began as access to publications, broadsheets, books and pamphlets detailing the latest scientific understanding became more widely disseminated. As Daniel Headrick states, 'public interest in natural philosophy grew during the 18th century, public lecture courses and the publication of popular texts opened up new roads to money and fame for amateurs and scientists who remained on the periphery of universities and academies'. There would seem to be little doubt that Joseph took advantage of the information open to him as he matured.[5]

Growing up on a small farm, Joseph would have learned the usual chores of a rural boy before becoming an apprentice blacksmith to his uncle, Thomas Powell. Later, this practical training was to pay dividends throughout his career. However, Joseph's keen mind and the potential he revealed was obviously noted, and his interest in the natural world would be honed at the Dissenting Academy of Llwyn Llwyd under the tutelage of its teacher David Price.[6]

FIGURE 2 Llwyn Llwyd House, Joseph and Howell's school
(photograph by author)

Large stone house, a former barn with an extension on the left-hand side of the door, a slate roof and wooden windows. The house stands in a large courtyard.

Llwyn Llwyd was merely a large barn near the village of Llanigon, a few miles from Trefecca. To modern eyes, the idea of a school in a barn does not sound very promising but nothing could be further from the truth.

The fact that Joseph and his brothers Thomas and Howell attended Llwyn Llwyd suggests that the family were not strongly attached to the Anglican church or became disinterested in it as time went by, or perhaps it was a form of education that could not be passed up. The schools were important in the education of bright young men due to the fact that the 1662 Uniformity act made adherence to the practices and sacraments of the Anglican church paramount to any academic or government career advancement. As Wales was becoming a mainly nonconformist nation, the dissenting academies were set up to combat this stifling religious ideology and open the world to its students. If any could afford it, this free-floating education system would allow them access to the Universities of Leyden, Utrecht, Glasgow or Edinburgh.

Joseph's teacher David Price was an ordained Methodist minister and a preacher at Maesyronnen church in Radnorshire (Powys) and he lived at Llanigon, teaching Classics, Greek, Hebrew, Latin, Theology, Medicine and Natural Philosophy (science) to its pupils. Though the school is listed as 'Migratory', meaning that it was not a full-time academy, Price must have instilled a great love of learning and erudition in his pupils, amongst whom were Joseph and his brothers Thomas and Howell. The great welsh hymn writer and Methodist William Williams of Pantycelyn was one of Price's pupils, studying medicine before opting for the career in the chapels he later embarked upon. Hugh Evans was another of Price's pupils who followed him into the Methodist system, eventually becoming an academy tutor in his own right. With such alumni around him, Joseph must have stood out as an interesting and bright character indeed.

According to M. H. Jones, Joseph seems to have been a quiet, introverted child who delighted in scientific experiment and observations of natural phenomena around him.[7] He appears to have been something of a tinkerer, enjoying making small instruments for varied purposes, although none now survive with the exception of a fire crane (a device for holding a pot or kettle over a fire) in the house at Tredustan, the village across the valley from Trefecca. Joseph also seems to have been taciturn and so lost in his own world that it appears the young agricultural workers who formed most of his circle of friends thought him inexplicable and, later, insane. How far we can actually take this portrait of Joseph as a young boy is open to interpretation, but much of it we can accept as his interests later in life were wide-ranging enough to encompass astronomy, mathematics, optics, navigation and instrument-making to name but a few, and perhaps many of these interests stemmed from his younger days.

Joseph must also have been an avid reader, although little evidence of his early education now survives. Joseph's later works reveal evidence of training in Latin and Greek, the mainstay of classical education at the time, so he must have applied himself well at Llwyn Llwyd and continued to develop such knowledge throughout his life. It is interesting to note that he taught Latin to his younger brother Howell at home. It is possible that Thomas Jones of Tredustan, then Sheriff of Breconshire may have opened his home library to Joseph so that he could amuse and educate the young prodigy. Jones paid a great deal of attention to Joseph and his influence would assist Joseph in his later career. Joseph's efforts at educating his younger brother Howell appear to have had a great influence on Howell's life as he became the only one of the Harris brothers to be accepted at St Mary's Hall College, Oxford, though he did not take up the position, mainly due to his differing ideas on doctrine and salvation and his lack of faith in the Anglican church system.

Joseph's demeanour, prowess and knowledge, as noted above, eventually brought him to the notice of the local establishment. We have already seen that the local squire, Thomas Jones of Tredustan (a large manor house just a mile away from Trefecca), saw much promise in the young man and probably encouraged him in his pursuits.

On 22 April 1715 a singular event that may have influenced Joseph's life took place across England and Wales: a total eclipse of the sun. If the weather was favourable, it would certainly have been watched by Joseph. From Trefecca the eclipse was only partial, but over 80 per cent of the sun was obscured. The general public across Britain were well informed of this event, as single-sided broadsheets printed by John Senex based

Figure 3 Tredustan Court (photograph by author)

Large eighteenth-century manor house of two storeys in an L-shape with the entrance at right-hand side. Built in local red sandstone with mullioned windows and extensive garden containing a well.

Figure 4 Library at Tredustan Court (photograph by author)

Large library with a baby grand piano in the foreground covered in a shawl flanked by small chairs with two reading lamps in the background. Behind the piano in the background are floor-to-ceiling bookshelves with volumes of books filling all shelves. A fireplace is at the left-hand side with an armchair close by and a table.

upon the calculations of the Astronomer Royal Edmond Halley and the publisher William Whiston were made available at a penny each. Although the weather reported by observers over England was variable, it is possible that Joseph would have glimpsed this phenomenon through the cloud gaps above his home.[8]

The Royal Society, including Halley, awaited reports on the eclipse from the 'curious publick'; it is possible that Joseph even at this early age may have prepared a brief account given his interests but, if so, any such report is now lost. If Joseph ever obtained one of these broadsheets it must have been serendipitous that he would eventually become apprenticed to the London cartographer and globe maker.[9]

FIGURE 5 Solar eclipse broadsheet by Senex (courtesy of the Royal Society)

Large single piece of paper with a map of the British Isles and the pathway of the 1715 solar eclipse across it from north-east to south-west and the area of greatest eclipse drawn in a darker hue.

By the time that Joseph became a young man, Thomas Jones had brought him to the attention of the local former MP, Roger Jones of Buckland, who not only encouraged him in his endeavours but gave Joseph introductions to the London Society of Instrument Makers in order that he may further his interests. That he was introduced to this society is perhaps telling in that Jones's tale of him tinkering with self-made instruments is probably true.[10]

Jones was no longer an MP after the 1722 election but maintained a strong interest in the society of his home county. It is also possible that Roger Jones may have taken an interest in Joseph as he had no children of his own and could see Joseph's potential just as much as Thomas Jones could. Roger Jones's letters of introduction opened a whole new world to Joseph as they would introduce him to the Astronomer Royal, Edmond Halley, and Colonel Samuel Shute, the Governor of the Massachusetts Bay Colony, amongst others. Encouraged by such attention and looking forward to a new life, at the age of twenty-one Joseph left the family home for London.[11]

It appears that his decision was motivated both by the glitter of scientific work and also by unrequited love. He had fallen for Anne, the daughter of Thomas Jones who had done much to further his career. However, her rejection of his advances probably made him determined to leave the area. As a local blacksmith, notwithstanding his excellent education, perhaps Joseph did not present the kind of gentleman that a squire's daughter ought to marry despite her father's interest in Joseph and his continued insistence that Joseph was an intelligent and worthy man. Despite residing in London, Joseph would maintain strong connections with Trefecca throughout his life.

Given this great start in life, how did Joseph fare in London?

2

LONDON AND APPRENTICESHIP

London in the early eighteenth century was fast becoming one of the largest cities in the world and taking its place as the centre of the burgeoning British Empire. From the perspective of a science historian, the capital in the eighteenth century was a place where the centralisation and interaction between scientific discoveries, mathematics, instrumentation, clockmaking, publishing, politics and finance were all advancing together under the auspices of men such as Isaac Newton, Robert Hooke, Christopher Wren, Edmond Halley, George Harrison, George Graham, Henry Cavendish, Joseph Priestley, Humphry Davy and so many others whilst its merchant marine and Royal Navy brought the riches of the planet to the door of the homeland.

Although romanticised by various people such as Samuel Johnson who quoted that 'if a man is tired of London, he is tired of life', the capital had its downsides too. Squalid, with closely packed buildings, disease-ridden from its burgeoning population and lack of sanitation, with a very high crime rate, the capital was a haven for murders, muggings and street brawls tumbling out of its many taverns; it was not until 1749 that a police force, the 'Bow Street Runners' was established, eventually to become the Metropolitan police. Punishments for crimes were generally harsh and the death penalty was handed out for even minor infringements of the law.

London was sprawling outward from the old city boundaries and creating suburbs that would suit the gentlemen and gentry of the day whilst retaining the squalid conditions and slums for the poor. Although the Romans, having occupied London over a period of four centuries, had established sewers for the city dwellers, these were overcome with the rapid growth in population and the Thames itself became an unhealthy and smelly open sewer. Sanitation was not a priority and the sights and smells of London in the early eighteenth century would have been overpowering for most. These problems were exacerbated by slaughterhouses along the banks that discharged offal into the river and ships emptying their bilges at the river banks as docks had not fully developed until the nineteenth century. Wharves lined the river for miles though it was only in 1799 that the West India Docks in what is now the docklands was built to ease the river trade.

Though the legal quays in the Pool of London at Billingsgate required ships to tie up there so they could be inspected for contraband, the river traffic was intense and crowded. Ships in the Pool of London were said to be so congested that it was possible to walk across the Thames merely by stepping from ship to ship as, in the main, ships merely tied themselves to posts on the bank or to each other.

The London of modern day bears little resemblance now to that of Joseph's time. In the early eighteenth century most of the city was built along the north bank of the Thames; south of the river was the borough of Southwark, then farmland sprinkled with villages; East Enders occupied a smattering of hamlets, and the location of the Greenwich observatory was in the royal deer hunting park which was a vast open space hemmed in by the Thames, Blackheath and the old Greenwich palace of the Tudors (Placentia). The Naval Hospital, which replaced the palace, eventually became the Royal Naval College but was not yet complete and would only be finished in 1742. Where Buckingham Palace now stands was an open field which had been used to grow mulberry trees to feed silkworms and, although there was a house there owned by the duke of Buckingham (hence the name), it did not become a royal residence until King George III purchased it for Queen Charlotte in 1761.

Not a single railway terminus was in evidence as trains were yet to be invented and travel around the city was mostly by foot or by cab or by using rowing boats. The only bridge spanning the Thames when Joseph arrived in the capital was the famed London Bridge, which connected the old City of London to what was then the separate borough of Southwark. The bridge was continually in a state of repair and disrepair, hence the children's nursery rhyme, though it has to be admitted that the bridge had been built in 1209 and much work had been done to keep it open. In 1670, traffic across the bridge was regulated to keep it to the left, thus introducing to Britain the idea of driving on the left. The buildings and shops that stood on the bridge were burned down several times in its history. In September 1725, a major fire destroyed all the houses and shops on the southern end of the bridge just as Joseph was going to the Caribbean. During his working life in the capital, London Bridge underwent extensive repairs and the final bridge with balustrades and fourteen alcoves for pedestrians to shelter in and a large central arch to allow shipping through was eventually completed in 1761. A second Thames crossing, Putney Bridge, was completed in 1729.

Trafalgar Square did not exist and the beautiful church of St Martin-in-the-Fields on the square's eastern side really was 'in the fields'. It was being rebuilt when Joseph arrived in London despite it surviving the great fire in 1666. It is almost unique in being a church where any and every faith is welcome. Burlington House in Piccadilly, now the home of the Royal Academy and the Royal Astronomical Society, had just finished undergoing renovation in the Palladian style and Devonshire House, the town house of the dukes of the same name, was always being added to and restructured until a fire destroyed it in 1733. Mayfair was expanding as was Marylebone. Slowly, villages such as Paddington and St Pancras and Bethnel Green and Shadwell became part of the expanding metropolis, and during Joseph's lifetime the capital swallowed up forty-six smaller villages, its two existing cities (Westminster and the City) and the borough of Southwark.

Dominating the entire skyline of the city was the huge cathedral of St Paul. Built after the great fire of London had destroyed the old one in 1666, the current cathedral was designed by Sir Christopher Wren and was finished in 1710. Standing on Ludgate hill, it was the tallest building in London until 1963. Joseph may have known about Ludgate's Welsh associations as, according to Geoffrey of Monmouth's *History of the Kings of Britain*, the pre-Roman Welsh king Lludd Map

Beli Mawr was the founder of London and was reputedly buried at Ludgate. Statues of King Lludd and his sons still stand in the porch of St Dunstan's in the West church in Fleet Street, a church Joseph would have passed many times as an apprentice.

By the time Joseph arrived in 1724, the city's population was estimated to be around 630,000 people amongst whom were up to 20,000 Welsh men and women living there along with itinerants such as drovers bringing their cattle, sheep and wares, the most important of which was Welsh wool, to the markets of Smithfield, in addition to the satellite markets of Billericay, Epping, Brentwood and Reigate.[1]

The Welsh were there in fluctuating numbers that probably never dropped below 10,000 due to the demand for workers, especially in the agricultural industry. Welsh women and girls, the 'merched y gerddi' tended gardens and fields in large numbers as the capital's demand for vegetables and fresh produce rose. Eventually they would be joined by their families and settled in the capital while retaining their Welsh identity. One of Joseph's uncles, Solomon Price had already moved to London and set up a tailoring business. When Joseph arrived, he would at least have some relatives he could depend upon and acquaintances such as the Morris brothers who founded the Cymmrodorion society and possibly the mathematician William Jones FRS who gave the world the mathematical notation Pi (π).[2]

There were some compensations for living in the capital. The composer George Frideric Handel, possibly the greatest composer in Britain at that time, was living in one of the new and fashionable houses of Mayfair (Brook St) and giving concerts at various venues including the King's theatre in Haymarket. Plays and comic operas could be seen at Lincoln's Inn Fields; the Queen's Theatre had opened in 1705 and was usually a packed house no matter what entertainment was offered. Covent Garden was also hosting music concerts and operas as was Drury Lane. Both Lincoln's Inn fields and Goodman's Inn Fields hosted free productions and Joseph may have made something of the performances during his time in London, although how much of this fare was derailed by the 1737 Licensing Act is open to question, preventing plays, musicals or other entertainments from lampooning public or political figures, a standard fare for performances of the time. Nonetheless, parody could always be found in the works of William Hogarth, who was not only an excellent artist, but was a satirist, cartoonist and social critic. For those looking for adventure, art, music and theatre, London had a lot to offer.

Information about current affairs, gossip and rumour abounded in the capital in its pubs and coffee houses. Educated Londoners were kept up to date with newspapers such the *Daily Courant* and the *London Gazette*. Periodicals such as the *Spectator* and *Tatler* were also available and Joseph no doubt made good use of these as he peppered his letters home with news of parliamentary ongoings and news on the monarchy. The Hanoverian King George I was on the throne having acceded to the Act of Succession after the death of William and Mary and, over Joseph's lifetime, two more Georges would ascend to the throne, George II and his grandson George III. Writing home to Trefecca over the years, we can see that Joseph kept his family informed of Bills passed by MPs, occurrences on the continent, contents of the King's speeches to Parliament and news on the various wars and conflicts that Britain was caught up with. He writes of the Prince of Wales's marriage to the King of Prussia's daughter, Augusta of Saxe-Gotha after a brief dalliance with, of all persons, Lady Diana Spencer. However, the prince could not afford the £100,000 dowry Earl Spencer, her father was insistent upon. (History was almost to repeat itself 240 years later.)

Such news and keen interest not just in the world around him but letters heavy with information on the royal family reveal Joseph to be a monarchist at heart. Although he befriended and worked with some of the highest-ranking ministers and MPs of his day, many of whom were at odds with the royal family, it would seem that Joseph was a steadfast admirer of them and his lifetime saw the reign of three of the Georges; he even drew a pension from George II.

Arriving here at the end of 1724, Joseph quickly began making friends and a name for himself amongst the precision-instrument makers and cartographers of his day. The twenty-year-old Joseph was apprenticed to the cartographer Senex and in fact moved into Senex's house in Fleet Street, writing home to his family that they should address correspondence there. The address was also close to the workshop of the instrument makers Thomas Heath and Jonathan Sisson in the Strand; it seems likely that he got to know them all on an informal basis as he seems to have done some design work that Heath later turned into workable instruments. M. H. Jones also claims that Joseph may have worked for Thomas Wright, of the instrument makers Wright and Cole who had premises in Fleet Street, but this early apprenticeship seems unlikely; although later Wright printed Joseph's *The Description and Use of the Globes, and the Orrery* as it contained a detailed description of the great orrery that Wright built. Any work that Joseph may have performed for any of these gentlemen must have been on a single or temporary basis, perhaps working together on various projects that are now lost in the mists of time.[3]

FIGURE 6 Orrery at former Royal Society Meeting place Crane Court (photograph by author)

A small model of a Solar System hanging over an arched entrance to Crane Court. The model is simply a globe with smaller globes fixed in place surrounding it.

It is difficult to tease out the threads of Joseph's involvement with these men as the world of scientific instrument makers was small and intertwined with astronomers, clockmakers, the Royal Society, Navy and Parliament. One constructor who strode across all these disciplines was Jonathan Sisson, the chief instrument maker to the Royal Greenwich observatory who would become the mathematical instrument maker to Frederick, the Prince of Wales in 1739. An apprentice of the horologist George Graham, Sisson started his own business just three years before Joseph arrived in London. Sisson would go on to become an instrument maker to the Navy and many of his designs and devices can be found in the Science Museum today. His astronomical quadrants and telescopes were famous, his eight-foot mural quadrant was commissioned by Halley for the Royal Greenwich observatory and even Pope Benedict XIV purchased instruments from Sisson for the Bologna observatory.

FIGURE 7 Halley's eight-foot quadrant (photograph by author)

Large wall-mounted quadrant (a quarter of a circle divided into 90 degrees) constructed of an iron lattice with a graduated quarter circle attached. Quadrant is in place at the Royal Greenwich Observatory with an attached sighting tube that can be moved up and down the quadrant to measure the altitude of stars above the horizon.

Sisson also made one of the first equatorial mountings for a telescope. His workshop would have been a veritable Aladdin's cave for a young man from rural Wales.

Joseph's boss John Senex was a renowned cartographer and engraver whose maps of the world were filled with additional details such as elevations and engravings, giving more information on places and geographical features. He had been the official geographer to Queen Anne and was eminent in the emerging field of geology. Senex was the first to produce a pocket atlas and in 1714 produced one of the first true atlases of England. Senex was heavily influenced by the French cartographer Guillame de L'isle whose maps of North America were highly sought. Senex himself was interested in the seaboard of what is now the United States of America and his maps of the Gulf of Mexico, Chesapeake Bay and the river basin of the Mississippi were excellent examples of his craft. In 1719 Senex produced a miniature version of John Ogilby's *Britannia*, a work of one hundred strip maps detailing the main roads and towns of Britain along with some details of each place. For his work and dedication to his country's interests, he was elected a Fellow of the Royal Society in 1728.

Senex was also a maker of terrestrial globes, each one being not just a conversation piece for nobles in some country house but being a valuable instrument through which basic principles of navigation, position, geography and astronomical calculation could be taught. Joseph's apprenticeship under Senex would pay later dividends as he used these principles in several of his later works.

Cartography was extremely important in the early eighteenth century; a growing industry as geographers and explorers began to open up the world and make exploratory missions into the New World, Africa and Asia. Additionally, great voyages of discovery were being made by Europeans navigating the Pacific and Indian oceans. New industries, eager merchants and the militia all needed accurate maps of their territories in order to understand their places on the globe and to exploit them.

One of the greatest mapmakers of his day, a contemporary and competitor of Senex, was Herman Moll, who had a shop and offices in the Strand, not far from Senex's Fleet Street store. In 1707 he began work on his *Atlas Geographicus*, which by 1717 had grown to five volumes and contained detailed representations of the world as known at the time. By 1710 he was producing small pocket globes and five years later, some thirty double-sided maps of the world in the form of a smaller atlas than that of *Geographicus*. Moll's maps were so famous that even Jonathan Swift mentioned them in his seminal work *Gulliver's Travels*. Moll was also close to the Royal Society through his friendship with Robert Hooke and the historian William Stukeley and no doubt enjoyed the company of the Australian explorer William Dampier who was the first person not only to describe some of the geography of Australia but to circumnavigate the world three times. Dampier was also the rescuer of the abandoned sailor Alexander Selkirk, who became the inspiration for Daniel Defoe's *Robinson Crusoe*.

Living so close to Moll, Joseph would have undoubtedly known him and become acquainted with his work and his eminent stature. Moll died in 1732 just as Joseph was arriving back from his second Caribbean voyage but, considering the associations that these men had with the scientists, explorers and literary figures of the day, one can see what a melting pot and meeting place London would have been for anyone keen on making a name for themselves, or, at least, contenting themselves that they were at the beating heart of Enlightenment wonders. It seems almost easy to have become involved if one had the inclination to do so and the intelligence to take advantage of the opportunities presented.

More importantly for Joseph's future endeavours were the letters of introduction to members of the Royal Society. The Royal Society of London was founded in 1660 and given a royal charter by King Charles II. It grew out of the meetings of learned men at Gresham College who wished to discuss the pertinent scientific and technological advances of their day. Including such luminaries as Robert Hooke, Christopher Wren, Isaac Newton, Edmond Halley, the society was one of the foremost think tanks of the Enlightenment. With its publication of *Philosophical Transactions*, its members and fellows could follow the latest ideas and reports on the wonders of their day.

The Society became instrumental in collating the scientific and technical knowledge emerging from the new scientific revolution. Its structure of regular meetings, publishing journals and subsidising scientific expeditions placed British science in the forefront of discovery. Under Queen Anne, the Society moved from Gresham College to new headquarters in Crane Court, just off Fleet Street, where the members were alongside an array of scientific instrument makers whose accuracy, precision and dedication to their craft were to make London the centre of a scientific industry that had few competitors.

Branches of astronomy, navigation, horology, cartography and the manufacture of precision instruments were overseen by the Society, which then distributed and applied such knowledge for specific purposes. In the early eighteenth century one of the most prominent problems was the accurate determination of longitude and the establishment of accurate distances within the Solar System to test Newton's theory of gravity. The natural philosophers of the age were concentrated in a relatively small enclave where the torch of renaissance thinking had passed from Europe to a nation now emerging on the world stage. Fellowship became open to foreign correspondents and a system of rewards, grants and medals established to provide incentives for natural philosophers to discover new theories underlying natural phenomena, materials, plants and animals in the natural world. The oldest of these is the Copley Medal and the honour of an annual keynote address, the Croonian Lecture was instituted in 1701.

Thanks to introductory letters from Roger Jones, Joseph was quickly befriended by the Astronomer Royal, Edmond Halley who took Joseph on a tour of the Greenwich observatory where he was shown the 'eight-foot quadrant' that Halley was using to measure the altitude of stars from the Greenwich meridian. Meeting Halley must have felt like a dream to Joseph. Not only was Halley the Astronomer Royal, but he was one of the greatest thinkers and intellects of his day. Savilian Professor of astronomy at Oxford, the secretary of the Royal Society, the man who mapped the southern sky; the man who became a naval captain of the ship *Paramore* undertaking voyages to study magnetic variation in the hope of using it to provide sailors with their longitude; the man who discovered Isaac Newton and paid for the production of *Principia*; the man who computed the first cometary orbit; the man who discovered the proper motion of stars; the man who invented the diving bell; a relatively humble man who did so much to promote the science and scientists of his day.

If it were not for Isaac Newton, it would probably be Edmond Halley that we would be holding in such esteem for promoting natural philosophy in the eighteenth century.

Another of the important persons Joseph encountered in London was the Welsh mathematician William Jones, the man who invented the term pi (π) to represent the ratio of a circle's circumference to its diameter. Jones was a very close friend of both Isaac Newton and Edmond Halley and was a Fellow of the Royal Society. Jones eventually became the vice president of the society in 1749. He may have taken an interest in Joseph and his later work on navigation as Jones had taught mathematics and navigation to the Royal Navy and even published *A New Compendium of The Whole Art of Navigation* in 1702. Jones split his time between London and his patron's home at Shirburn Castle, Oxford, the estate of Thomas Parker, the Earl of Macclesfield and may have encountered Joseph in London on many occasions. They certainly shared an acquaintance with the Morris Brothers of Anglesey (whom we shall encounter later), both men knowing Richard Morris well. Jones had also written many mathematical tracts and letters for the Royal Society dealing with astronomical problems and the true shape of the Earth, problems that Joseph would later deal with during his sea voyages.

Over the next few months during his time as an apprentice, Joseph made, refined and improved instruments for use in navigation. He constructed a new form of azimuth compass able to read magnetic variation to one degree and also designed a more accurate forestaff for navigational purposes, which was constructed by Thomas Heath. The azimuth compass was an important contribution as such instruments could be used by navigators to align the compass to a given celestial body such as the pole star or the Sun. The reading from such an alignment would then give true north and any deviation could be read as magnetic variation. In addition to these tasks, Joseph also busied himself by honing his cartographic skills and making the most of Senex's knowledge of geography and the problems of longitude.[4]

Joseph, Halley and the heavens

Halley and Senex already had a very close working relationship as Senex had printed the maps of the track of the 1715 solar eclipse calculated by Halley and also produced maps of both the northern and southern hemispheres based on Halley's surveys of the sky. As we have seen above, Joseph's letters of introduction paved the way for him to work with Halley. Their meeting must have impressed the Astronomer Royal, who saw something in the young man that singled him out for future greatness. It was no doubt Joseph's attention to detail and his earnest endeavour to be of some use that impressed Halley, adding to the influence of the information passed on by Roger Jones of Buckland, that would give Halley the idea of sending Joseph abroad to test the accuracy of his surveys.

FIGURE 8 Portrait of Edmond Halley (courtesy of the Royal Society)

The Astronomer Edmond Halley in profile wearing a white wig and brown jacket. Underneath is a white shirt with a white cravat signifying that he has the religious title of Reverend, which was common to University graduates of his day. Halley is about 40 years old. Painted for the National Portrait gallery.

In 1724 Senex published the Cambridge astronomer, William Whiston's *Calculations of Solar Eclipses*, a booklet recommended to him by Halley that contained a compendium of eclipses for the next few decades and would be used by Joseph the following year. The predictability of celestial events and the unchanging nature of the stars must have appealed to Joseph. That cartography could be applied to the celestial globe as well as the terrestrial one was a fascination that was to put his name literally on the map. Through this association with the Astronomer Royal committed to his cartography apprenticeship, Joseph was probably encouraged to prepare a map of the southern stars under Senex's guidance.

However, there is some ambiguity over when such a chart was first drawn up. The hindsight of posterity is not helped by the fact that celestial cartography in the early eighteenth century in London was a tale of woe and infighting with a short and volatile history.

Senex's celestial maps use Halley's star charts of the southern hemisphere and unlicensed copies of the northern hemisphere which were taken (stolen) from John Flamsteed's unpublished charts of the sky for his *Atlas Britannica Ceolestis*. Whilst first Astronomer Royal, John Flamsteed was a touchy and irritable character who did not get on well with Halley, who had been his former assistant. Additionally, Flamsteed could not stand Isaac Newton who was now the president of the Royal Society and a frequent visitor badgering Flamsteed for the results of his work. Flamsteed forbade any use to be made of his star charts produced over many years of work at the Greenwich Observatory, seeing them as his personal property.

In fact, the first publication of Flamsteed's *Atlas* was controversial, as much of it was stolen by Halley and Newton, both pillars of the Royal Society who were insistent on their publication in order to help navigators, which was, in essence, why the Greenwich Observatory had been founded. Enraged by the recalcitrance of Flamsteed, they descended on the Greenwich observatory and took as much of the work as they could find. The bootlegged charts were given to John Senex who quickly printed the first single-sheet star maps based on them.[5]

The larger work of preparing the entire atlas took a little longer.

Flamsteed resisted so much that when the unauthorised version of the *Atlas* was published by the Royal Society, Flamsteed purchased almost every copy and publicly burned them. In the event, the *Atlas Britannica Coelestis* did not see the light of day until Flamsteed's wife published it in 1729, ten years after her husband's death.

Against this fractious background, the old enemy Halley became Astronomer Royal after Flamsteed's death in 1719 and wanted to produce accurate astronomical charts as quickly as possible. Having access to the as yet unfinished *Atlas Coelestis*, his friendship with Senex ensured that general star maps quickly became available, including maps of the southern hemisphere that Halley had charted from St Helena in the southern Atlantic between 1676 and 1677. His work detailing 341 southern stars was published as the *Catalogus Stellarum Australium* in 1679 and it is this work that Senex and Harris later improved upon. We shall return to this theme in Chapter 4.[6]

The maps later became very important as they were affordable, accurate and withstood much use. Maps of the night sky were relatively few and far between as most star catalogues were in published tomes such as Johann Bayer's 1600 *Uranometria*, Tycho Brahe's *Star catalogues*, the former Astronomer Royal John Flamsteed's *Atlas Coelestis* or in other published forms that were not easily accessible. They were also very expensive, and few individual astronomers could afford them. Senex began to change that with the printing of accurate and affordable star maps that were part of the growing scientific understanding demanded by the educated public. One of Joseph's lasting legacies is his application to stellar cartography and influence upon this particular form of publication.[7]

With the hindsight of history, this situation reveals a recognition of his talents that such detailed and accurate work was entrusted to someone who was apparently still an apprentice. It appears obvious that Joseph was a careful and gifted cartographer. His expertise was no doubt noted by members of the Royal Society; and certainly by others in his circle. It seems incredible that Joseph would be picked for such a task unless his expertise was already legendary and his skills so trusted that Edmond Halley personally selected him and probably supported him financially on his 1725 Caribbean expedition as Joseph would have been removed from his apprenticeship under Senex.

It is evident that Joseph's apprenticeship and association with the close-knit band of instrument and mapmakers ensured that his name and expertise would be recognised by leading men of science of his day. Joseph was trained by some of the best craftsmen in London in a trade that was to become internationally pre-eminent by the end of the century. His improvement of the azimuth compass for the Navy and other, now lost, inventions reveal the depth of insight, accuracy and diligence demonstrated by Joseph. Alongside his co-workers, he focused on precision instrument making and mapping, making practical application to the problems of his day.

Joseph's dedication to accurate cartography and celestial mapping must have been remarkable for it to be brought to the attention of the Astronomer Royal and others who then supported Joseph in his near future travels. Additionally, the concentration of clockmakers and instrument specialists in Fleet Street and the Strand ensured standards of excellence and competitive interplay whilst introducing Joseph to some of the most eminent instrument makers and cartographers of his time such as George Graham and James Short. The instrument makers also had their work promoted by the Royal Society with many of the best, including Senex, becoming Fellows.[8]

Considering his apprenticeship and associations, it is unsurprising that Joseph would make himself as useful as possible in the service of the sciences of his day. Some of his most far-reaching contributions were to be made during his Caribbean voyages and it is to these that we now turn our attention.

3

THE CARIBBEAN VOYAGES

Joseph's expertise in the production and use of instruments and charts was noted by Halley and others and so were his observational and mathematical abilities; even at this early stage it is possible that his early tinkering and potential designs of instruments also impressed. It is not possible to ascertain if some of Joseph's instruments such as the forestaff and improved azimuth compass were tested on these voyages as just six months were to pass between his arriving in London and his first voyage. It is possible that Joseph's description and uses of some of these, including the instruments of Heath, Sisson and George Graham, were subsequently tested on his second voyage to the Caribbean. Nevertheless, his voyages to Vera Cruz in modern day Mexico between 1725 and 1727 and another to Jamaica between 1730 and 1732 established Joseph as a person of trust and sound judgement in addition to his application to the task ahead. As we have seen previously, it is possible that he was financially supported on his first voyage by the Astronomer Royal, Edmond Halley to whom he sent at least one letter detailing his observations.[1]

Seafaring in the early eighteenth century was a dangerous undertaking laden with perils that ranged from merely cramped conditions to mutiny, injury, diseases such as scurvy and the poor food, the main part of which consisted of salted beef or pork, fish, ale and ship's biscuit. The quality of the food often became problematic as poor ventilation and vermin on board the ships made inroads into it. What Joseph may have made of this meagre fare is not recorded, though he may have dined with the officers who did have a more varied diet. Coupled to these immediate problems was estrangement from home for months or even years, piracy and the daily problems of weather and unpredictable seas. In addition, there was also the problem of accurate navigation as there was no sure way to measure longitude at sea and would not be whilst Joseph experienced his two voyages. In short, there was no guarantee that any ship venturing across the Atlantic Ocean would actually reach its destination or even return home. Having one of the largest navies afloat, and a merchant fleet almost equally as large in number, was no assurance of success.

Whether these problems came to Joseph's mind is not known. Perhaps as a young man undertaking a foreign adventure, he was excited and thrilled to be playing a part in such a scientific exploit. He wrote home to Trefecca that he had received orders to board ship and, possibly feeling a bit homesick, he mentions Thomas Jones and Anne of Tredustan perhaps hoping to impress her and to thank his mentor for the opportunities he had presented.[2]

The first Caribbean voyage

On 15 June 1725, two ships set off from the UK bound for Vera Cruz in Mexico: the *Prince Frederick* and the *Spotswood* sailing under the flag of the South Sea Company. Joseph writes of going aboard a ship loaded with £300,000 of goods and crewed by 250 men. Though Joseph was impressed by the value of the cargo and the size of the ship, never having been at sea before, all was not well behind the scenes, and he was unknowingly sailing into an international incident.

In 1713, Queen Anne and King Philip V of Spain signed the Treaty of Utrecht, which ended Britain's involvement in the wars of the Spanish Succession that had taken place since 1701, eventually ending in 1714. Under the agreement the Spanish gave the British authorities an *Asiento*, or licence to conduct the slave trade with Spanish colonies in the Americas. The contracts to carry out this work was given to the South Sea Company and an intensive marketing campaign was launched to persuade people of wealth to buy shares in the company, leading in 1720 to the collapse of the market and the resultant recession based on the 'South Sea Bubble'. This was the first financial crash in history, which began when a British joint stock company called the South Sea Company was founded in 1711 by an Act of Parliament. It was a public and private partnership that was designed as a way of consolidating, controlling and reducing the national debt and to help Britain increase its trade and profits in the Americas. Under the Treaty of Utrecht, the company was granted a trading monopoly in the region, dealing in the trading of African slaves to the Spanish and Portuguese Empires. The slave trade had proved immensely profitable in the previous two centuries so there was huge public confidence in the scheme, as many expected slave profits to increase dramatically, especially when the War of the Spanish Succession came to an end and trade could begin in earnest.

The South Sea Company began by offering those who bought stocks an incredible 6 per cent interest. However, when the War of the Spanish Succession came to an end in 1713, the expected trade explosion did not happen. Instead, Spain only allowed Britain a limited amount of trade and even took a percentage of the profits. Spain also taxed the importation of slaves and put strict limits on the numbers of ships Britain could send for general trade, which ended up being a single ship per year. This was unlikely to generate anywhere close to the profit that the South Sea Company needed to sustain it. To promote confidence, the South Sea Company bought up much of the national debt and released stocks at inflated prices until the inevitable happened: the stock was seen to be overvalued and the shares, once trading at £1,000 per share, plummeted to £124 with massive financial loss as the shares became almost worthless. Investors were ruined and a subsequent parliamentary enquiry uncovered bribery and corruption at the highest levels.

Unfortunately, the South Sea Company was seen by the British government as another arm of its imperialist intentions and the company was basically let loose to trade as it pleased. Maximisation of profit was more important to the company than any agreement and soon the South Sea Company violated the treaty.

Under the terms of the *asiento*, 4,800 slaves a year could be brought to the colonies in America along with a stipulated portion of goods. The colonial inhabitants of Vera Cruz, Cartagena, Havana, Portbello and Buenos Aires looked forward to receiving not just the slaves but the goods brought by the South Sea Company as they were difficult to obtain locally. This situation gave rise to a ready market for European goods amongst the settlers leading to the agreements largely being ignored and excess materials traded. Over a few years the South Sea Company continued to stretch and break the treaty repeatedly by bringing more slaves and goods than the *asiento* allowed. The company even went beyond its remit and actively bribed local officials. In the case of the *Prince Frederick* in which Joseph was sailing, they 'gifted' the local Spanish and Mexican Viceroys with a 'sword garnished with diamonds and an exquisite musical clock'.[3]

Bringing the two ships loaded with goods was incendiary and it was no surprise that trouble between Spain and Britain broke out over the amount of cargo being carried. The *Prince Frederick* was seized by local officials despite the bribes and detained for several years; its commander, Captain Williams eventually dying in central America whilst still under investigation. The *Spotswood* was released and, after the conclusion of its trade, went home.[4]

Breaking the terms of the Treaty of Utrecht in this way was bound to have later repercussions. Things eventually came to a head whilst Joseph was on his second voyage (and living in Jamaica), when a Spanish privateer caught a South Sea Company ship, the *Rebecca*, illegally trading off the coast of Cuba. The captain of the ship, the Welshman Robert Jenkins, had his ear cut off by the privateers as a reprimand, which almost a decade later led to the 'war of Jenkins ear', lasting from 1739 to 1748. Clearly it reveals the extent of British illegal operations but also the power of the South Sea Company in instigating a conflict many years after the event in pursuit of its profits.

It seems likely that Joseph sailed out on the *Prince Frederick* but returned via the *Spotswood*. Although he seems impressed that the *Prince Frederick* was loaded with a great deal of trading materials and by their subsequent value, he mentions nothing of slaves or slavery, recording nothing of this human cargo: this traffic was *de rigueur* for Britain at the time and slavery apparently did not cause him too much worry, as it was a normal part of British life in the early eighteenth century.

The bustle, heat and very alien nature of the Spanish colony must have made quite an impression on a young man who just a few months before had never ventured further than a few miles from home. No letters from Joseph have survived, so what his impressions were and what he could have communicated to his family are completely unknown. Although no doubt enjoying his experience, he still had work to do and, whilst at Vera Cruz, Joseph determined the latitude and longitude of the town and carried out observations of eclipses and the night skies with the four-foot quadrant that he mentions in his letter to Halley. Quadrants are instruments used to measure the altitude of the stars above the horizon and mark their positions for inclusion in a celestial map. Joseph also notes that a telescope could be fixed to the quadrant, which he did for the 1727 solar eclipse, though he neglects to tell us exactly what the aperture and focal length of the instrument were.[5]

It is after his observations of this eclipse that he was able to determine the longitude of Vera Cruz by calculating the differences in timing between Flamsteed's tables of London times based on the Greenwich meridian and the timings obtained locally from the clocks on this expedition. His calculations revealed Vera Cruz to be 19° 12m North and 90° 30m West.[6]

A paper detailing his results entitled 'Astronomical Observations at Vera Cruz' was published in the *Philosophical Transactions* in 1728. Joseph's contribution to these observations was a collection of data regarding a partial eclipse of the sun on 11 March 1727, and observations of a lunar eclipse on 29 September 1726, which was partly clouded out. This report was read at the society's meeting by Edmond Halley, who noted the accurate determinations of latitude and longitude obtained by the young observer, which corrected the position of the city from its former map location. Beyond these brief details, not much is communicated in this paper. Joseph does not include any observations of the stars in his report, so it can be presumed that this work was ongoing.[7]

Within the paper are two remarks on instrumentation that are simultaneously arresting and annoying to a reader far removed in time.

The quadrant of 'four-foot radius' that he mentions is an interesting instrument. Large and cumbersome, it needed time and patience to set up. His letter to Halley does not detail if this was a mural quadrant, one requiring fixing to a vertical wall, or whether it would be part of a free-standing frame. Joseph also notes that he observed the position of the Moon across the solar disk during the solar eclipse by means of a telescope attached to the quadrant. What was the aperture and focal length of the telescope? What filter did he use if this was used to directly observe the Sun? Did he follow the practice of blackening the telescope with charcoal or soot to provide some direct relief for his eye, or was the image projected in some way? Sadly, we will probably never know.

It is obvious that he was in possession of some very fine instrumentation, though whether these were loaned to him by Halley, or made for the voyage by Sisson or Coggs, can only be guessed at from the paucity of the information he conveyed. It is known that Halley encouraged him to complete as much as he could of any outstanding work on the southern stars, so no doubt Joseph was engaged in this activity in addition to noting any other phenomena he encountered.

Within this correspondence Joseph points out another problem he was assessing, the impediment of magnetic variation on compasses, lamenting 'I always found the best observations we could make when compared together differed so much that we could not depend on them to much than three or four degrees and sometime half a point of the compass'. He spent the greater part of 1726–7 plotting this magnetic variation around the Caribbean, following up on Halley's work of almost fifty years previously and no doubt using the *Spotswood* or another South Seas Company ship which he fails to mention. Magnetic variation was a major concern for course-plotting and Joseph would go on to make further reference to this problem which now prompted something in him to think of the broader astronomical and navigational significance of his work on his return home. It was a problem he returned to on his second voyage.[8]

Nevertheless, the importance of this determination of longitude and latitude at Vera Cruz becomes clear when one considers that this was the era in which the need for accurate longitude was paramount to sailors. In 1714 the Merchants and Seamen's Petition resulted in Parliament passing the Longitude Act of 1714, which heralded the importance of determining longitude for both the Royal Navy and the merchant ships of the UK. Although the act supplied a series of rewards for anyone who could accurately determine longitude it is unlikely that Joseph was motivated by this and used methods, calculations and tables already drawn up by others, notable amongst whom were John Flamsteed, the first Astronomer Royal and of course Edmond Halley himself who succeeded Flamsteed.[9]

The application of new instruments and accurate global positioning was a major driving force of astronomical and navigational science during Joseph's lifetime. His observations and mathematical dexterity coupled to his skill as a cartographer enabled him to make a vital contribution to this field even though it seems that he was careful to remain in the background of scientific publication. It appears that his modesty forbade him to inscribe his name on the instruments or maps that he fashioned, unlike many of his contemporaries.

Even his reports were read to the Royal Society by others and his name added to them by the person delivering the report. Why Joseph wanted to remain in the background as a mere voice in the choir is somewhat of a mystery.

Upon his arrival home, Joseph no doubt completed his apprenticeship and continued to make friendships and acquaintances amongst the Royal Society who must have acknowledged him as a trusted gentlemen and a possible future Fellow of the society. Between his arrival home and leaving for his second voyage, Joseph returned to Senex and completed the 'Harris' star maps, the book *A Treatise on Navigation* (1730) and the text for *The Description and Use of the Globes, and the Orrery* (1731). His heart was set upon teaching at this point and it is possible that he was a private tutor in mathematics to various important families; but, if so, he does not record this.

What did interest him, of course, was navigation. He must have spent a lot of time with the officers and navigators of the *Spotswood* as it plied around the Caribbean and no doubt with those on board the *Prince Frederick* on the outbound journey. His magnetic variation measurements required as accurate a position as could then be obtained and undoubtedly Joseph learned all he needed to know from these navigators before adding his own astronomical and mathematical knowledge, improving on their methods and observing their errors. The experience and application of his understanding in this field between 1725 and 1727 shone through in *A Treatise of Navigation* as we shall shortly see.

The second Caribbean voyage

By the end of the decade, Joseph was considered by many in positions of influence within the Royal Society to be a trusted hand and he was asked, probably by Halley and communicated by the secretary Cromwell Mortimer, to conduct further observations and experiments when he was chosen to accompany Colin Campbell to Jamaica in 1730.

On 14 October 1730, Joseph set sail to the town of Black River where Campbell planned to establish a permanent observatory equipped with a transit circle and a mural arc of four-foot radius made by Heath and Sisson; instruments that would enable them to determine a local line of meridian in order to make accurate observations and timings. According to James Bradley, the two men were to 'improve astronomy and promote other parts of natural philosophy in that island'. Joseph was therefore engaged as a highly trained assistant to Campbell, a typical employment for someone of talent and ingenuity but lacking the wealth or social connections of the 'gentlemen' of the day. It was an acknowledgement of his standing in the scientific community and his known attention to detail. No doubt he would have taken the 'Harris Maps' with him in order to study the southern sky and complete any outstanding differences.[10]

Whilst in Jamaica, Joseph noted in May 1731 that he had not received letters from his family and, as the sailing season was now over, he could expect none until the following year as the sailing season would once again open between October and February. This seasonal closure was probably due to the fact that the high summer in the Caribbean is hurricane season and ships were in grave danger if caught up in one of these storms. Additionally, part of the family's silence was due to the problem of timing: getting letters to Joseph on the other side of the Atlantic at the right season and the fact that his father had died in March of that year. Perhaps the family did not want to upset him with such news when he was in no position to return and comfort them.[11]

Joseph notes in this letter that he has had 'severe seasoning', probably some sort of tropical disease such as malaria or yellow fever (he neglects to say which one), but he assures them that he has now recovered. Despite his being enchanted by Jamaica with its large estates, abundant woods and good weather, he admonishes them not to go abroad as 'no other country enjoys so many blessings as does England'![12]

Joseph's work in Jamaica as a scientific assistant included the testing of clocks, recording magnetic variation, and making astronomical observations that repeated the observations of Jovian satellite eclipses noted in 1726 by Bradley. This work was vital in establishing time differences between as many geographical locations as possible to facilitate the calculation of longitude. The positions of the satellites of Jupiter – Io, Europa, Ganymede and Callisto – had been observed since the Italian astronomer Galileo Galilei discovered them in 1610 and their association with the problem of longitude had been discussed by Galileo, who invented a helmet with an inbuilt telescope to observe them from a ship. This device was impractical but the idea of timing the movement, transits and eclipses of the satellites from a fixed position so as to discover longitude by the differences in timing from another location became a practical way of determining longitude as long as Jupiter was above the horizon. Joseph's task here demonstrated the method could work, but it was probably useless aboard a heaving ship at sea and doubly useless if one could not see the planet.

Joseph continued to assist Campbell in observing the sky from their latitude of 18 degrees north (which latitude Joseph established) and acquiring precise celestial positions of the stars in order to complete Halley's *Catalogue of Southern Constellations*. Although Campbell never finished this task, Joseph's dedication to it was already evident as, of course, he and John Senex had produced large-scale celestial maps just a few years previously.

Campbell and Joseph also set up a clock with an isochronal pendulum made by George Graham in order to determine its accuracy in the heat of the tropics and for an experiment to test the pendula of clocks at different latitudes. The clock had been set originally for sidereal rather than solar rate in London by using a meridian transit of the star *Altair* (α Aquilae) and comparisons were made with other timepieces to note any differences. Sidereal rates are based on the movement of the stars as the Earth rotates. Although we normally consider the day to be a full twenty-four hours, the sidereal rate of rotation where a star transits a fixed point daily differs by four minutes, being twenty-three hours and fifty-six minutes. This is the true rotation rate of the Earth and so a clock fixed by a sidereal timing should produce a different transit timing if used anywhere else on Earth. The local Longitude could then be determined from the difference in timings. The clock arrived with instructions from Graham on how it was to be set up and what prior experiments had been performed with it in London.[13]

The purpose of this experiment was also to test Isaac Newton's theory that the force of gravity may diminish with distance from the equator. Campbell found that the heat lengthened the pendulum and the time slowed at a rate of nine seconds a day; there was an additional slowing possibly due to other factors. Both observers noted that the clock lost fifty-four minutes and twenty-one seconds after twenty-six days. When they had adjusted the clocks, given appropriate attention to their running, length of pendulums and other factors, they were faced with a small dilemma that could only be understood if the force of gravity itself was affecting the clocks. Their observations gave insight into an eighteenth-century problem known as the 'Figure of the Earth'.

Isaac Newton and Christiaan Huygens had both postulated in the seventeenth century that, as the Earth is rotating, the true shape of the Earth should be an oblate sphere where the distance around the equator would be greater than that of a circle from pole to pole. This should have an effect upon the gravity of the planet where the force of gravity would differ slightly from the equator to the poles and subsequently have a slight effect upon the swing of the clock's pendulum depending on its global position. It was eventually determined that the equatorial and polar differences are only thirteen miles.

These precise measurements performed by Campbell and Joseph were communicated in person by Joseph in July 1732 to James Bradley FRS (who would become the third Astronomer Royal), who compared them with readings at Uppsala, Sweden taken by Anders Celsius and those of Pierre-Louis Moreau de Maupertuis's French expedition to Lapland which took one of Graham's clocks and a zenith sector, again constructed by Graham. Bradley was then able to suggest some adjustments to pendulum clocks for different latitudes and tropical heat, and the combined measurements enabled Maupertuis to determine the true shape of the Earth as an oblate spheroid. These observations once again introduced Joseph to some of the great scientists of his day and reveals not just the internationalism of science during the Enlightenment, despite the political differences of their countries, but the circles that Joseph was comfortably moving in.

During the two years of this expedition, Joseph travelled around the Caribbean, continuing his work on magnetic variation, noting local offset near Florida and Havana by using small portable compasses. Back at Jamaica, Joseph drew attention to a phenomenon he independently discovered, *diurnal magnetic variation*, noting that compasses altered as much as two degrees a day, principally in the forenoon. The phenomenon had been recorded by George Graham in London in 1722, but this was the first time it was recorded at another location on Earth. It is testimony to Joseph's meticulous work that he noticed it at all since the variation was no more than one or two degrees. Joseph admitted having no idea as to the cause of this occurrence; it has since been found to be caused by ionospheric electric currents generated by the solar wind, though some variations are associated with different solar and lunar effects.

Once again, Joseph's meticulous observation and attention to detail would reveal phenomena that would, centuries later, explain such things as the Aurora Borealis as hinted at in 1741 by Anders Celsius.[14]

Most of the instruments used by Campbell and Harris at Jamaica were sold to Alexander Macfarlane, a merchant, postmaster-general and judge at Jamaica who wished to improve upon the astronomical observations of Campbell and communicate them to the Royal Society. In the event, Macfarlane made only two known observations: that of a lunar eclipse on 2 November 1743 and a transit of the planet Mercury three days later. The instruments were later bequeathed, upon Macfarlane's death, to Glasgow University, which built an observatory equipped with these instruments in 1760. They included a horizontal reflecting zenith sector, a four-foot mural arc, a five-foot transit telescope, a portable zenith sector and a one-month regulator clock, in addition to miscellaneous lenses, micro-meters, compasses, an astrolabe and a camera obscura. It is testament to Joseph's knowledge and skill that he was not only able to use each of these instruments, but to set them up and keep them in excellent order.[15]

Joseph perhaps took a quiet personal pride in his contribution to these vital scientific activities, though, infuriatingly, he claims so little credit for his work. Although there is barely any further information beyond that found in the *Philosophical Transactions of the Royal Society*, Joseph's writings only illuminated the extent of his activities in a broad sense; he must nonetheless have been an enthusiastic and gifted young scientist.

He did communicate the weariness and lack of appreciation of the endeavours both he and Campbell felt when he told Bradley, 'Mr Campbell hinted that if something should be said in favour of what we have done, it would be some sanction to him in that part of the world where astronomy is in but little esteem'. It had obviously been a long and arduous expedition carried out in difficult circumstances.[16]

This voyage would be the last time that Joseph travelled abroad. Graham notes that he was taken ill and had returned from the Jamaican expedition earlier than planned. Whatever this illness was, as we have seen above, is not recorded, but malaria once again remains a possibility given that he refers vaguely to 'fatigue or gout' in several letters and mentions it again during his observations of Venus from Trefecca. Although this is a record of an event almost thirty years afterwards, chronic forms of malaria have recurring effects leading to drowsiness or tiredness many years later. Of course, various infections and fevers were common in the eighteenth century and it is possible that Joseph merely suffered from many of the common ailments of his time rather than a tropical disease. Once again, his humility has left us guessing at what could have ailed him.[17]

This humility perhaps prevented him from accepting a greater reward. It appears that, despite his application to his craft, he was not at this stage invited to become a Fellow of the Royal Society or there is a possibility that he turned down the privilege. Potential fellows would have been approached by their referees before a submission to the Society. From what we know of Joseph's humility it is possible that he was approached with the offer of a fellowship but turned it down. A search of the archives shows that there is no record in the meeting minutes at the Royal Society in his lifetime that his name was ever forwarded for consideration. His later contributions to the *Philosophical Transactions* attribute the epithet of 'gentleman' to him, without letters after his name as with other authors. Indeed, during his sea voyages his work on observations and the production and use of instruments was as a subsidiary assistant to the main observer/leader who was a Fellow of the Society.

Arriving back in Britain in the summer of 1732, Joseph sends a letter home to Trefecca with a large quantity of goods: their value and what they were is lost as Joseph does not detail them and we only find out about this bounty from a letter his brother Thomas sends to Trefecca. In a letter to his brother Howell, Joseph mentions that he saw John Conduitt MP in London upon his return and he was going down to Hampshire to stay with him at Cranbury Park for a while. How Joseph actually came to know Conduitt and why he was offered such a privilege is unknown, but this association was to prove a blessing to Joseph as Conduitt at this time was master of the Royal Mint and was perhaps looking for someone able and exacting to work at the mint. We shall return to this later as Joseph did not immediately take up a position there.[18]

Instead, Joseph struck out in a new direction.

Upon his arrival home, he became a teacher of mathematics at Gosfield Hall, near Braintree in Essex, the home of John Knight MP.

Turned down for the position of Navigation teacher at Portsmouth some years before, perhaps the experience had soured Joseph's approach in returning to this mode of employment. However, having a background in so many diverse subjects, teaching maths to the young daughter of John Knight and having a permanent base at Gosfield Hall obviously appealed to him and allowed him some settlement where he had easy access to London and contact with his brothers and mother. Many letters to the family during this time are franked in London as perhaps the post was more reliable from the capital.[19]

Arriving at Gosfield Hall in August 1733, Joseph no doubt took to the family and his duties as a teacher in his typical manner. He does not mention his charges or pupils as Knight's only son had died in 1727, so beside Knight's remaining daughter he may have been a teacher of the local children and used the Hall as a base. How often he taught is unknown as many of his letters between 1733 and 1735, when he finally took up residence at the Mint, put him in London for most of his time but do not give many clues as to what he was doing or how he was employed, though a single letter from Joseph written from Gosfield Hall in September 1735 shows that he was still in some measure involved with the household.

FIGURE 9 Gosfield Hall (photograph by author)
Imposing white façade of the front of Gosfield Hall with Georgian windows and archways in Palladian style. The roof is penetrated by Mansard windows.

It is instructive to find that Knight's widow Anne (nee Craggs) in 1737 went on to marry Robert Nugent MP, a figure that Joseph dealt with in some part throughout his career at the Mint as Nugent (or Robert Craggs-Nugent as he became known, after adopting his wife's name and fortune) served as lord of the Treasury between 1754 and 1759. It is possible that Joseph and Anne may have attended the occasional party thrown by Mrs Nugent in London over the next few years though what he may have thought of her is not recorded. Satirical pieces of the time show Anne Nugent to be 'fat and ugly' with her and Robert constantly quarrelling in private. She died in 1756 and was buried at Gosfield Hall.[20]

This interlude probably allowed Joseph to build connections with prominent men in the worlds of politics and science that would assist him in the years ahead and eventually elevate his advice, intelligence and application to the very heights of power.

4

NAVIGATION, CARTOGRAPHY AND MATHEMATICS

Returning to London in 1727 after his first taste of Caribbean adventure, Joseph probably resumed his apprenticeship with Senex and, once this was completed, began seeking work as a teacher of mathematics (as he later styled himself) and navigation. In 1730 he produced his self-financed work *A Treatise of Navigation*. No doubt he had taken a keen interest in the subject whilst sailing the Caribbean and in 1733 after his second voyage he applied to become a teacher of navigation at Portsmouth. He was unsuccessful as other candidates were 'supported by great interests'. Contemplating the broad interests he had nurtured under Senex, Joseph collated his thoughts and ideas on globes and orreries in addition to his ideas on navigation and celestial cartography and went on to produce some of the best publicly available star maps of the eighteenth century. Together these works mark him out as a remarkable and able young man.

Cartography and the 'Harris Maps'

We touched upon Joseph's contributions to celestial cartography in the last chapter but it may be instructive to look at it in greater detail. Through his association with Halley, he was probably encouraged to prepare a map of the southern stars under Senex's guidance, a map that came into general use by 1728. However, there is some ambiguity over when such a chart was first drawn up. Senex's celestial maps use Halley's star charts taken from John Flamsteed's unpublished charts of the sky for his *Atlas Britannica Ceolestis*. Halley became Astronomer Royal in 1719 and wanted to produce accurate astronomical charts as quickly as possible. His friendship with Senex ensured that star maps quickly became available, including maps of the southern hemisphere as Halley had charted these from St Helena in the southern Atlantic between 1676 and 1677. Senex and Halley introduced a version of Halley's southern star map in 1678 and his work detailing 341 southern stars was published as the *Catalogus Stellarum Australium* in 1679. It is this work, updated in 1690, that Senex and Harris improved upon.[1]

It is quite possible that the improved maps for the southern celestial hemisphere were originally arranged by Joseph between January and June 1725, before he set off for his first Caribbean voyage with the proviso that his observations would improve the accuracy of the southern star maps. The 1725 expedition is noteworthy. This is very soon after Joseph's arrival in London and his introduction to Halley. Halley impressed upon Joseph the need for accurate measurements of his southern stars before this expedition was undertaken so it remains possible that Halley wanted to test the accuracy of the new chart of the southern stars. This is conjectural but further evidence may be found in the fact that on his second Caribbean expedition he was tasked by Colin Campbell to check the stellar positions of Halley's *Catalogue of Southern Constellations*. Campbell failed to do so but Joseph competed the work. It would appear that perhaps a large part of his early apprenticeship with Senex was spent compiling the southern hemisphere chart, eventually printed in 1728 after his return from the Caribbean, a situation that had surprising consequences.

Though the celestial maps sold by Senex utilising the above observations came into common use, it is Harris's name that has been associated with them over time. On the map of southern stars, it is noted that the map was created by Joseph Harris utilising the stars observed by Edmond Halley at St Helena. No such citation is on the northern charts which draw upon the *Atlas Coelestis* that John Flamsteed had produced. So, how did these maps become known together as Harris's maps? Did Joseph continue to engrave, draw and improve them between 1725 and

1728 so that, in common parlance, both maps became associated with him rather than with Halley or Senex? As Deborah Warner states, 'the internal relationships between scientists, cartographers, publishers and dealers were often so complex as to obscure the specific contribution of each'.[2]

Such a careful cartographer as Senex would have wanted to have the utmost accuracy. Perhaps it is this dedication that is reflected in the

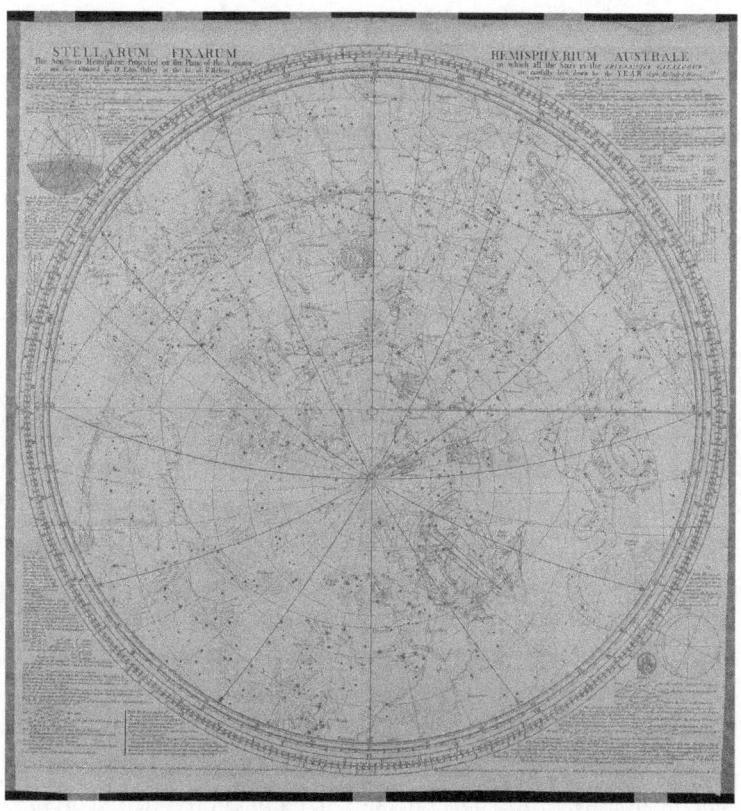

FIGURE 10 Harris's southern sky map (courtesy of the Royal Greenwich Observatory)

Joseph Harris's map of the southern sky with constellations drawn on it in classical design of mythological creatures and stars showing different brightnesses. Joseph's name appears at the top right-hand corner of the map above the date.

fact that Senex produced the maps, but the cartographer (certainly of the southern map) was known to be Joseph Harris; both maps eventually becoming ascribed to him by astronomers (despite the northern map being produced before Joseph arrived in London) and his precise work known to the Royal Society. Just as we use an author's name on a publication but ignore the publisher, do the 'Harris Maps' reveal their later reprint and authorship, with Senex relegated to the role of printer?

Although Joseph does not communicate it, perhaps his work during his first Caribbean expedition enabled him to obtain more accurate positions for some of the southern stars and this resulted in the 1728 publication of the Senex/Harris map of the southern stars upon his return – an update on the 1725 work. Did they work together on producing a new northern chart at this time so that both hemispheres could be printed at the same time and updated? Is it possible that Senex already had the manuscript for Halley's *Atlas Britannica Coelestis* (released in 1729) and wanted to reduce the work to something simple and immediately useable so that his northern and southern hemispheres were reviewed and redrawn by Joseph?

It may not be possible to prove this, given our distance in time and the lack of records. Nonetheless, the importance of these maps is again highlighted by Warner: 'even after 1729 when the authorized version of Flamsteed's atlas appeared, because of the convenience of the single sheet maps and their relatively low cost, Senex's maps continued to be popular with both astronomers and navigators'.[3]

The 'Harris Maps' remained in general and professional use for almost one hundred years and were used by the noted astronomer William Herschel in his star sweeps and catalogues of deep sky objects. Indeed, Herschel was using the Harris Map when he discovered the planet Uranus in 1781. Today some of Harris's maps are in the curatorship of the Royal Museums Greenwich, highlighting the importance of accurate celestial mapping in both determining longitude and for astronomical use in the eighteenth century. Perhaps Joseph would have been flattered that his craftsmanship would be so recognised; his maps being used by one of the most pre-eminent observatories in the world at that time. If he was, he kept such emotions to himself.[4]

Joseph Harris: Scientist, Artisan, Assay Master

A Treatise of Navigation

In the spring of 1730 Joseph released his work, *A Treatise of Navigation*, which started beautifully as a shot across the bows of navigators under British employ, stating that 'middle latitude sailing is grossly erroneous and that the common method of keeping reckonings in meridional distance is grossly false'. Joseph then goes on to lay out several problems in computing the bearing of a ship in different circumstances and draws upon astronomical objects and their known latitudes and timings as a more accurate way of fixing position than dead reckoning. The book is addressed to Viscount Torrington, Lord Archibald Hamilton, Sir William Young and the lords commissioners of the lord high admiral of Great Britain in addition to others. Joseph obviously hoped to make a favourable impression with these gentlemen but ensured that the work was dedicated to Edmond Halley who had done so much to assist Joseph in his endeavours.[5]

Within *Treatise*, he discusses his invention and uses of the Azimuth Compass, a new design of forestaff which was simpler and more comfortable to use and which was produced en masse by Thomas Wright, John Coggs and Thomas Heath of Fleet Street whom he mentions in the text.[6]

Joseph's intention in writing the book was to save lives at sea by challenging current navigational techniques and replacing them with a more mathematical approach coupled to the use of improved instrumentation. We need to remember that this was written almost forty years before chronometers were used to judge longitude at sea based upon accurate computations and timings at Greenwich. As he states: 'I have endeavoured to clear the theory of navigation from errors that have crept into it and to make such improvements in the practical part as based upon my own experiences.'[7]

How many navigators of the time could actually follow his improvements is open to interpretation as the first chapters deal with plain and spherical trigonometry, disciplines that Joseph may have felt to have been in relatively short supply amongst navigators in the eighteenth century as many, especially naval officers, owed their positions to family connections or wealth, though of course there was a navigational school at Portsmouth where such maths disciplines were no doubt part of the course. Rear Admiral Mark Kerr makes the point that 'It is evident that the essential principles of sound navigation were already in place by the time he put quill to paper, what was missing was the technology to exploit them to the full.' Joseph's hope was that his design and use of better instrumentation might make an important difference to the accuracy and safety of ships at sea. The text and the persons to whom it was addressed perhaps reveal a change of attitude toward navigation on the Navy's part to do away with dead reckoning and an acceptance of the expertise of the London instrument makers attempting to assist the navy with more accurate techniques of guiding ships to their destinations and return.[8]

At the back of the Naval mind was the tragedy of Admiral of the Fleet Sir Cloudesley Shovell, who in 1707 led his entire battle fleet to destruction on the rocks of the Scilly Isles. A young midshipman had challenged the admiral on his positioning, only to be immediately hanged from the yardarm for disputing an officer, though this tale has allegedly been discredited. Nevertheless, the battle fleet, one after another, foundered on the rocks and over 2,000 men lost their lives. Against this background, any help in navigating a ship may have been preferable.

Treatise goes on to discuss the uses of Mercator maps whilst advocating better plain charts and the use of trigonometry in fixing the best navigational position based on course, speed, experience and faith in their own abilities. He returns in the text to the Mercator projections as being of great use at sea if they are accurately drawn and published. He also warns against not taking sea currents into account when calculating the heading and position of a ship. The problems he then demonstrates show once again the breadth of Joseph's mathematical ability, perhaps harking back to his days at Llwyn Llwyd under David Price's tutelage but undoubtedly reinforced by his own application, experience and learning.

In order to assist navigators compute quickly, *Treatise* contains detailed tables of the Solar position, the stars fixed by the charts of Flamsteed and Halley, their use of logarithms coupled with sines and tan tables that could be immediately referenced by a navigator at sea and a ship's position quickly fixed, even if the resultant position may not be accurate by today's standards. Interestingly, he also includes some materials on the use of slide rules in computation. Reviewing the book from a twentieth-century perspective, Rear Admiral Mark Kerr states:

> Harris writes with some considerable experience of deep sea voyaging. This practical background is combined with a solid understanding of those principles you would expect a 'teacher of the mathematicks' to know and an evident determination to improve the practice of navigation by 'persons of ordinary capacity'. The result is a book which positively seeps best practice, well described.

Subscribers to the book included Thomas Jones of Tredustan, the poet Alexander Pope and Anne Knight, the wife of John Knight MP.[9]

Useful though the book may have been, it did not achieve his future purpose: to obtain employment at the Naval school in Portsmouth when he returned from his second Caribbean voyage. Joseph laments to his brother Howell in March 1733 that, although he put in for the position, there were others in line for the job who 'had more favourable connections than I'. Nonetheless, the book became a valued contribution to navigational techniques at the time and remained so for many years, being reprinted several times.[10]

The Description and Use of the Globes, and the Orrery

As we saw in the last chapter, Joseph accepted the position of assistant to Colin Campbell's Jamaican expedition in late 1730. Previously, upon his return to London in 1727, Joseph collated his knowledge of navigation and mathematics and, drawing upon his experiences with Senex (who was a globe maker as well as printer), in early 1730 he left with his publisher a second book known as *The Description and Use of the Globes, and the Orrery*. Within this work, he taught navigational techniques in an early attempt to produce a workable system of establishing an observer's position on the Earth. The book was released in 1731.

A great part of the book is not Joseph's at all but, rather, was written by the unrelated John Harris as a navigational treatise on the use of globes; this publication first appearing in 1703. Known as *The Description and Uses of the Celestial and Terrestrial Globes; and of Collins's Pocket-Quadrant*, Joseph enlarged the book by adding a first section on the Solar System and an addendum on using Thomas Wright's famous orrery, constructed for the Naval Academy at Portsmouth.[11]

However, Joseph applied his astronomical knowledge to practical problems as can be gauged by reading the *Description*: the first section is an introduction to the Solar System and the stars, filling the first thirty pages of the book; he then revisits celestial positioning many times throughout. The later sections about globes and orreries present positional astronomy, the motions and positions of the planets in an accessible manner and enable the reader to grapple with navigational and positional problems.

What is less appreciated is why Joseph added to this known work and why he focused on the Solar System in particular.

To our modern interpretations the term 'Solar System' is an obvious description of the Sun and its family of planets. Mercury, Venus, Earth, Mars, Jupiter and Saturn were familiar to many people (Uranus, Neptune and the Kuiper belt had yet to be discovered) watching the skies, but the acceptance of a Sun-centred system was, remarkably, still in its infancy. Up until the start of the eighteenth century the idea of a heliocentric system was known as the *Copernican Hypothesis*.[12]

Although British artwork, outlining the new description of the Sun's position and that of the planets, existed as early as Thomas Digges's 1576 illustration in his work *A Perfit Description of the Celestiall Orbes*, it was not until 1696, just thirty-three years before Joseph's description, that William Whiston's *A New Theory of the Earth* opened with a representation of the *Systema Solare* that displays an actual planetary system as accepted today. The term 'Solar System' was not to be introduced into the English language until 1704.[13]

The idea of a Solar System with the Sun at centre and planets in orbit around it was therefore relatively new to the public; it is this intriguing concept with which Joseph wanted to familiarise his readers.

Joseph Harris: Scientist, Artisan, Assay Master

To examine how novel the concept was, the first English writer who specifically mentions a *Solar System* in its usual meaning was Savilian Professor of Astronomy David Gregory (1659–1708), who introduced the term in his *Astronomiae physicae* in 1702 and *Geometricae Elementa* of 1704. It is also possible that the philosopher John Locke first coined the term 'Solar System' in his *Elements of Natural Philosophy*; though it had also been used in Latin and German in 1705 by Johann Gabriel Doppelmayr.[14]

To put an introduction specifically outlining the Solar System into a known text that would be of use to navigators is a significant step on Joseph's part that goes beyond the mechanical wonders of the orrery he later describes. He immediately saw the potential that the planets can be used as a natural device open to mathematical interpretation to show that geometry can be used to study their basic positions.

He also states that the planets are globular and opaque to light, rendering them solid bodies, or at least, bodies that reflect the sunlight rather than have light in themselves. He also uses proofs to show why Venus and Mercury show phases like that of the Moon, as bodies that orbit the Sun inside the orbit of Earth his geometry reveals the different phases are due to position.[15]

One of the more interesting problems he discusses is the distance to the planets from the Sun, and this in an age when the Earth–Sun distance was unknown. He uses trigonometry to assess their distances using their position angle in relation to the Earth and Sun and produces a table on page 28 of *Descriptions* in which he provides the orbital characteristics in time (years and fractions of a year) and their distances. Incredibly, his results are within 10 per cent of modern values using these methods. He also informs the reader of the relative sizes of each planet, again computed in miles, and, again, these are within a few per cent of their modern value.[16]

His introduction to the Solar System therefore is of great interest as he shows that calculation and mathematics can reveal much about their basic positions, distances and relative sizes. He makes the Solar System a known quantity, subject to physical laws and open to mathematical interpretation. Though such methods had been used for over a century since Johannes Kepler formulated his laws of planetary motion, Joseph brings an accessibility to his Solar System, showing that the Sun's family of planets was a real, measurable entity, not a hypothesis.[17] Furthermore, Joseph describes comets as bodies that

> must be very hard and durable bodies, else they could not bear the vast heat that some of them, when they are in their *Perihelia*, (their closest approach the Sun) receive from the Sun, without being utterly consumed. The great Comet which appeared in the year 1680, was within ⅙ part of the Sun's diameter (about 230,000km) from his surface; and therefore its heat must be prodigiously intense beyond imagination. And when it is at its greatest distance from the Sun, the cold must be as rigid.

He also hints that due to their extreme eccentricity of orbit, they must come from a great distance from the Sun and notes that some have retrograde motions and high inclinations. Demonstrating a knowledge that is in keeping with our modern studies, he considers them as virtually sold bodies that follow the same physical laws as the planets. It must be kept in mind that it was only a few years earlier, in 1705, that Halley computed the orbit of the comet now ascribed to him.[18]

It is interesting to note that, amongst the descriptions of the stars of the Milky Way and the constellations, Joseph records that the planets of the Solar System are 'subject to the same laws of motion with our earth, and as some of them not only equal, but vastly exceed it in magnitude, it is not unreasonable to suppose that they are all habitable worlds'. He reveals a familiarity with common scientific speculation of his day and continues in this vein to extend the idea of habitable worlds to planetary systems around other suns also. Though such speculations appear strange now, the idea of extraterrestrial life across the Solar System was an acceptable premise promoted by the publication in 1686 of Bernard Bouvier de Fontenelle's *Conversations on the Plurality of Habitable Worlds*, in 1686. Such free conjecture was only reined in (briefly) by the 1854 essay, *Of the Plurality of Worlds* by William Whewell, who ruled out the existence of such life on theological grounds. Joseph stayed within a familiar theoretical backdrop of his day by ascribing the existence of such life to the magnificence and wisdom of a creator.[19]

He leaves this interesting philosophical discussion to concentrate on the purpose of his text. One of the problems he highlights (problem XXXV) is concerned with measuring the longitude of stars from the first point of Aries and then combining an earlier problem (VII) of global positioning to find an approximate terrestrial longitude. The key to the problem was the standardisation of a point of reference from which the rising times of particular stars visible across the globe could then be matched. Joseph used mathematical methods to calculate longitude by revealing it to be a problem of stellar position from a known meridian, a method that was recommended by the Royal Society and Greenwich Observers. He was keenly aware that such methods demanded continuous clear skies, something of a rarity in Britain, and his second Caribbean voyage would incorporate early experiments with the use of clocks as potential tools in the hunt for longitude.[20]

Joseph put his advice to good use by making reference to his calculations of the setting times of fixed stars from the Lizard, Cornwall in 1729. He wrote up these exercises for the publication of his next book, *A Treatise of Navigation* in the spring of 1730. The book attracted wide approval and was reviewed by the Astronomer Royal, Edmond Halley who added a dedication in the preface. In an early biographical essay on Joseph, the historian M. H. Jones makes the mistake of attributing the fixing of the planet Neptune's position in 1729, to Harris from the Lizard in his *Treatise of Navigation*. Uranus was not discovered until 1781 and Neptune in 1846. Clearly, this information is incorrect.[21]

Further scientific contributions: navigation, observations and demonstrations

On his return from Jamaica, Joseph applied for a job teaching navigation at the naval school at Portsmouth, but did not get the post. He then privately tutored mathematics in the household of John Knight MP and occasionally taught at the Royal Society's museum at Crane Court, just off Fleet Street where he reacquainted himself with many of the clockmakers and instrument makers of the day, who had shops nearby.

Being in such a position, he was ideally situated to pick up news on the latest scientific and instrumental advances and his intellectual circle widened to include James Short, the famous telescope maker, and also some of London's greatest clockmakers, including Thomas Mudge and the mechanical genius John Harrison, who would eventually prove the use of clocks in determining longitude accurately. On 2 May 1733, George Graham observed a solar eclipse from Fleet Street with a ten-foot focal length telescope. The report was subsequently communicated to the Royal Society and it is likely that Joseph would have been interested in this phenomenon and watched the event as he was in London at the time.[22]

His occasional teaching duties at Crane Court, Fleet Street, the new home of the Royal Society since 1710, included demonstrations of the use of scientific and navigational instruments. During this period, he came into close contact with the instrument maker John Hadley, vice president of the Royal Society with whom he shared an interest. Hadley had refined the octant and sextant for navigation and had also built the first reflecting telescope in an appropriate design for astronomical use. Hadley also sat on the committee appointed by the Royal Society to assess the instruments obtained by Edmond Halley for the Royal Greenwich Observatory and was aware of Joseph's voyages and skills.[23]

Over this same period Joseph continued a close correspondence with his parents and his brother Howell, and in 1735 came down to Trefecca with the express purpose of persuading Howell to matriculate at Oxford. Their letters reveal the chaotic state of Howell's mind, as he was willing to leave the college and become an itinerant Methodist preacher. Joseph, however, took a dim view of his brother's preaching activities, a tension that would continue throughout their lives. Joseph's training in maths and science and his fraternity with members of the Royal Society show that he favoured a natural view of the world rather than one that included the hand of a vengeful and capricious God that Howell espoused.

Given what he had written about the hand of a creator in the habitation of the planets in 1729, it is more than likely that Joseph's distaste for Howell's preaching activities may have arisen due to popular perception that such 'itinerants' were outside the church – the Anglican denomination refusing to accept them as priests. It is probably concern for his brother, who may have been stepping away from a lucrative career as a country parson, that led Joseph to think the negative thoughts Bennett attributes to him. Whatever the position, he encouraged Howell to stay at Oxford and get his degree in theology, even sending him clothes in early 1736 to last the term.[24]

Despite the distractions of sibling life back in Wales, Joseph continued to apply himself to his mathematical and nautical work until in 1736 he was appointed as the Deputy Assay Master of the Royal Mint in the Tower of London. Work at the Mint was exacting and the hours long, but Joseph managed to maintain his scientific interests and correspond with his mother and siblings, travelling to Trefecca as often as he could. In between these activities he still found time to attend meetings of the Royal Society and continued to cultivate his friendships there.

In 1739, he demonstrated the uses of globes at a meeting of the Royal Society and later wrote up an account for the *Philosophical Transactions*. Joseph showed how to make an horary circle, detailing the hours of the day on a dial, and intersecting both poles of a globe to become moveable about an axis so as to calculate times from a fixed meridian to the east or west and also to make calculable allowances for deviations in latitude. He then 'made some contrivances to shew the effects of the Earth's motions and also adapted it for shewing how the vicissitudes of day and night, and the alteration of their lengths, are really occasioned by the motion of the earth'.[25]

The report reveals that Joseph still dabbled in instrument making and other applications and discloses his ongoing interest in using instrumentation to simplify calculations or to teach difficult concepts, as the *Transactions* go on to give an account of a simple machine demonstrating how the annual motion of the earth in its orbit causes the apparent change of the sun's declination, all without the expense of making an orrery.[26]

Joseph was highly skilled, attentive and a stickler for detail and these qualities shine through in his books, his observations and no doubt in correspondence and conversations with the learned men of his day. Although his contributions may be overlooked or subsumed into the larger scientific understanding and applications of his time, Joseph had a finger on the pulse of the nation in a fashion shared by so many other persons of his time. His job as assay master at the Royal Mint would bring him into the higher political and scientific circles of his day, yet he humbly refrained from pushing himself forward or demanding recognition. It is to these contributions that we now turn.

5

ASSAY MASTER AT THE MINT

The Royal Mint is the oldest established company in the UK that has been owned by the Crown Treasury from the seventeenth century to today. In Joseph's day, the Royal Mint was in the Tower of London, which was established in 1279 under the rule of Edward I. The three main administrative roles at the Mint were Comptroller, Master and Warden and, as increasing sophistication and counterfeiting of coins over the years made inroads into the value of and trust in coinage, the post of Assay Master was added by the fifteenth century.

Minting of coins in the UK has a long history. Possibly the first recognised coins in circulation were those used by Celtic tribes crossing the English Channel in the second century BC and then the Roman invasion brought the industry of minting coins to Britain shortly after their 43 BC conquest. By the time the Romans left in the fourth century AD, a mint had been established in London for some time. However, this closed after their departure and no coins were minted until the establishment of Saxon kingdoms in the sixth and seventh centuries. By 1279 the varied mints in operation were centralised by Edward I and the Tower of London Mint came under crown control. Although a few exterior mints struggled on until the time of Henry VIII, the dissolution of the monasteries saw coin production brought under the almost total control of the Tower.

In 1603 the Union of Scotland and English monarchies under James I saw an imbalance in the value of coinage between the two countries where a Scottish mark was worth 13.5 pence and an English mark worth 80 pence. To make up the shortfall, supplementary coins made from lead were minted as tokens of exchange. In order to ensure parity, the Mint eventually hired the 1st Baron Harrington under licence to produce copper coins and in 1672 the Tower Mint took over total control of copper coinage.

In 1630, the government began to import Spanish silver bullion to increase the standard and value of money coins in the UK. To aid in this endeavour, additional mint branches were opened including one at Aberystwyth castle in Wales. During the English Civil War various mints were established by Charles I who moved around the country and needed money to retain his armies. Upon Charles's defeat and the establishment of the Commonwealth of England under Oliver Cromwell, coins were now minted with English inscriptions rather than Latin ones. At the end of the civil war, the Mint returned to its main headquarters in the Tower of London and the auxiliary mints across the country were closed down. The government then invited the French designer and engineer Pierre Blondeau of the Paris Mint to modernise British coinage and, despite his efforts at milling coins, Cromwell's death saw a return to the hammering of coins in the minting process. Upon the recall of King Charles II, Blondeau returned and introduced a milled edge to British coinage that reduced the problems of clipping and weight restriction.

The glorious revolution that ousted James II from power saw the mint controlled completely by Parliament, a situation that has held to this day. However, due to clippage, counterfeiting and a need for standardisation, all coins from the late seventeenth century onward had to be generally withdrawn and replaced as the value of precious metals had surpassed the face value of the coin. To rectify the situation, King William III ordered a great recoinage and under the watchful eye of Sir Isaac Newton, the master of the mint, all coins were now to be milled rather than hammered. This led to the introduction of new machinery to the Tower Mint which had to be tested, examined and kept in a good state of repair to ensure a high degree of accuracy in the minting of coins.[1]

Employees of the mint therefore had a lot of work to do and had to ensure weight, accuracy of measurement, accounting and a knowledge of the processes at work in the Mint. Considering his background in cartography, astronomy and instrument making, how did Joseph end up working at the Royal Mint?

It appears from correspondence that Joseph had been in touch with John Conduitt MP, master of the mint after the death of Newton and Isaac Newton's nephew through marriage to his niece (and adopted daughter) Catherine. Conduitt became the executor of Newton's estate after the great man's death in 1727. As Newton lived with the Conduitts at Cranbury Park in Hampshire during the latter part of his life, it can be imagined that they would have had some intriguing conversations that ranged from the end of the world, superstitions and legends, through to the value of money and the political problems of the day. Conduitt's parliamentary record is notable as he introduced the Witchcraft Act of 1735, which made it illegal to accuse others of practising witchcraft, leading to the abolition of the hunting and burning of witches. It is obvious that as members of parliament (Newton also served as an MP in two terms between 1689 and 1702) there was close association between the mint masters and the issues of their day.

Joseph had stayed with Conduitt at Cranbury Park for a month in 1732 and, even though Conduitt died in 1737, he had already recommended Joseph to Richard Arundel MP who was the new master. Many of the mint's masters were Royal Society men who would have known of Joseph's attention to detail and meticulous record keeping in addition to his ability as a precision instrument maker. As we have noted above, they were also Members of Parliament who had ready two-way access to important economic work through their deputies at the Mint and their ministerial positions. Joseph would prove an ideal choice in an exacting branch of civic work that demanded scientific application and accuracy.[2]

In 1696, Isaac Newton became master of the mint and took his position seriously enough to ensure that the Royal Mint would have no rival. At the time, forgeries accounted for almost 10 per cent of the coins in circulation, the Spanish Bullion coins from earlier that century now had a value that far surpassed the face value of the coins, whilst coins from counterfeiters and auxiliary mints were still in circulation but with various values. Across the UK, the difference in coin value between their metallic value and face value became an economic burden and many small 'mints' were set up to make money tokens that could make up the difference in coin value. The system was a mess.

The great recoinage introduced by King William III in 1696 could not have come at a better time, ensuring a fresh start to the Royal Mint and passing into law that the use of counterfeiting equipment would result in charges of high treason, a sentence that carried the death penalty by hanging, drawing and quartering. If the forger was female, they were merely burned at the stake. During Joseph's tenure, any woman accused of forgery was deported.

The recoinage resulted in all coins across the UK being withdrawn from circulation and the Royal Mint put into production a new series of coins. To ensure the country had enough coins in circulation, the Tower Mint authorised 'satellite' mints in Bristol, Chester, Exeter, Norwich and York. The comptroller of the Chester Mint, recommended by Isaac Newton, was none other than Joseph's future friend and mentor, Edmond Halley, who spent two years at the mint supervising coin production. Once the new coins were in circulation, these auxiliary mints were closed down and production once more moved to the Tower of London.[3]

In 1707, the Act of Union with Scotland closed the last remaining mint in Scotland and the Royal Mint took over production of coins for that country. This led to increased pressure on the Royal Mint, which was working in wooden buildings at the Tower in unhealthy, cramped and dangerous conditions as the smelting of the base metals needed for coins took place under stuffy and badly ventilated conditions. The metal was cast into fillets of the breadth and little more than the thickness of the intended coin and were then reduced to their correct size by a rolling mill operated by horses tramping round a cellar below. Thus people, animals and the process of producing coins were squeezed into small spaces making the work noisy, hard, smelly and unhealthy from the noxious gases and poisons such as lead, zinc and antimony used in coin production or released during the smelting process.

Security surrounding the mint was very tight. This is hardly surprising given the amounts of precious metals being trafficked within the walls but, additionally, the tower was still functioning as a prison.

Several prominent Jacobites were held there such as the Irish sympathiser George Kelly, who escaped the tower just before Joseph's arrival. Simon Fraser, the 11th Lord Lovat, was executed there for treason in 1747 and Flora MacDonald was held there between 1746 and 1747 for assisting Bonnie Prince Charlie evade capture after the battle of Culloden. In 1716, William Maxwell the 5th Earl of Nisdale famously escaped the night before his execution by dressing as his wife's maid and even Sir Robert Walpole, a future prime minister, had been imprisoned in the tower in 1712 on charges of corruption. Imagine if that were still the case!

Mint workers were not allowed to fraternise with other workers in the tower such as the warders or visitors. Mint staff living at the tower were kept separate from the rest of the tower's inhabitants and their movements were strictly monitored. The mint itself was spread throughout the tower with wooden temporary buildings scattered around, and more permanent buildings in the outer ward which became known as Mint Street. Anyone working at different levels at the mint had to go through distinctive and oppressive levels of security, perhaps many times a day. Despite having relatively luxurious apartments, the top echelons of the mint staff must have felt like prisoners themselves on occasion.

This situation existed in Joseph's time as it was not until 1810 that the mint moved to new purpose-built headquarters on Tower Hill outside the Tower of London. In addition to the above, the Tower still had a tidal moat filled by the Thames. This moat was the source of several epidemics amongst the Tower staff as the filthy water, filled with offal and excrement, gave rise to disease. It was not until 1843 that the moat was drained on the orders of the Duke of Wellington, who had become constable of the Tower.

Additionally, the tower still retained a menagerie of animals, the London zoo of its day, which was not closed until 1835, long after Joseph had died. What Joseph and Anne (Anne Jones of Tredustan whom he married in 1736) may have thought of the proximity of some dangerous wild animals can only be guessed at, especially as some had in the past killed visitors and keepers alike. Whilst Joseph and Anne lived in the tower, a visitor to the menagerie, Mary Jenkinson, had the flesh of her arm 'torn from the bone' after attempting to stroke one of the lions. She died hours later. It could be surmised that Joseph and his family kept a respectful distance.[4]

Against this background, Joseph took up employment at the mint on Lady Day 1736 (25 March), eventually becoming deputy assay master in 1737 under Hopton Haynes the King's assay master. For Joseph, 1736 was a jubilee year: a new, secure job and, after many years of secret courtship, letters and arrangements by Howell, he now went on to marry Anne Jones, the daughter of squire Thomas Jones of Tredustan, who had probably brought Joseph to the attention of society all the way back in 1724. Together, the newlyweds moved into the village of Hackney and, years later, moved into the Tower as was the fashion for mint workers and their families.

Hopton Haynes had become assay master in 1696 at the same time as Newton was appointed as Master of the Mint and he worked there for forty years before he was allowed to appoint Joseph in his place. Haynes and Newton got on extremely well as they were both Unitarians and had a dedication to the Mint that went beyond mere service; indeed, Newton had recommended him for the post due to his 'integrity, sobriety, good humour and readiness in business'. Their concentration on monometallism and the gold standard became bywords for precision and accuracy at the mint and in 1749 Haynes retired on a full pension but continued to be an auditor for the government Exchequer (a role that would fall to Joseph in his turn) until his death in November of that year at the age of 77.[5]

Haynes and Joseph must have had an excellent working relationship that endured through many problems, with both men dedicated to hard work and government service. Interestingly, Haynes had written a volume in 1702 about the recoinage and the corruption of hammered money leading to the reform of coinage and the need for proper milling. Joseph no doubt noted this and may have paid some reference to it in his later *Essay upon Money and Coins*.

It is a measure of the trust in him as a person, in addition to his abilities, that he was appointed, as the statutes of the mint required assayers to be 'professionally competent and regarded with confidence inside and outside the mint'. As seen above, there had been an extensive recoinage under Sir Isaac Newton ten years prior due to the 1707 Act of Union with Scotland and the need for a unified currency. Newton had established the gold standard for coinage, but there was no accurate standard for copper or silver monies. Joseph was appointed as an officer in charge of the accurate measuring of the gold standard and began to work on establishing a standard for money already in circulation and to institute a new standard for minted coins.[6]

Joseph worked alongside some excellent craftsmen who designed several new coins of gold and silver during the reign of King George II. Following his monometallic ideals, new guinea coins (a guinea is the equivalent of £1 and 1 shilling) made of gold, from five guineas down to half-guineas minted with the King's face on one side and a shield surmounted with a crown on the obverse. This gold was obtained from the British East India Company up until 1739. A proposal by Reverend Peter Vallavine to put a graining or 'dotted' edge to the coins and make the lettering larger as a way of avoiding clipping and counterfeiting was adopted by the Mint and overseen by Joseph.

Coins with the new designs were made by two men, John Croker and John Sigismund Tanner. Croker and Joseph worked briefly together as, by 1739, Croker was an old man and was too unwell to produce the dies for the coins. Tanner was his apprentice and took over from him and became the Mint's chief engraver following Croker's death in 1741. Tanner outlived Joseph by eleven years but they must have had a productive relationship, going through several recoinages including one in 1757. Like Joseph, Tanner also received a royal pension and lived in the Tower in the Chief Engraver's house. Together Joseph and Tanner produced what is thought to be the first known 'proof set' of coins for the mint in 1746: the Crown, Half-Crown, Shilling and Sixpence and all were made of silver. Up until 1971, when Britain went decimal, a sixpence was known as a 'tanner' in possible deference to its die maker, John Sigismund Tanner. The sixpence was thought for many years, before and after this coinage, to have magical properties as a changeling woman (one who could turn herself into animals) could only be shot with a bullet made from a sixpence. What Joseph would have made of this myth can only be guessed at.⁷

The work was extremely demanding and took its toll on many of its workers. Hopton Haynes was taken ill in the spring of 1748 and Joseph, as deputy, was left to carry on the tasks alone whilst training an assayer who could take over should he in turn fall ill. The record keeping by the weighers and tellers under the jurisdiction of the assay master had to be absolutely accurate and this would probably prove to be a stressful process. As the ingots of silver and gold came into the mint, their value and weight were recorded and passed on to the comptroller and warden before the metals were taken to be smelted and turned into coins. At the end of that process the coins were weighed so that the incoming and outgoing weights of the precious metals tallied and no pilfering or loss of materials was to be found. Any discrepancy would result in an investigation and subsequent loss of trust in the mint and its workers, a situation that would have massive ramifications. Thankfully, no such situation arose during Joseph's tenure.

By the autumn of 1748 Joseph became the King's Assay Master as Haynes retired at the age of 76. Joseph and his wife Anne took charge of a larger house in the Tower of London between the Beauchamp and Devereux towers. Their proximity to the Tower from their original residence in Hackney would have been very useful on busy occasions, such as that back in 1745 when British privateers captured two French ships loaded with gold and silver treasure with a value of over £800,000 at eighteenth-century rates. Joseph records that he was under enormous pressure reducing the materials and producing coins from them. Some of these coins are still in existence amongst collectors and are prominently stamped 'Lima' after the city in Peru, whence the treasure was taken.

Possibly to combat the unhealthy atmosphere at the Tower, and the proximity of the open sewer that was the river Thames, Joseph and Anne also had an alternative residence at Grove Street in Hackney close to the open fields of the east end where Victoria Park now stands, about an hour's walk from the Tower. From 1760 they also had access to Place House in Lewisham, south of the Thames where they could get away from the hustle and bustle of central London.

FIGURE 11 Devereux Tower, Tower of London (photograph by author)
Devereux Tower stands at the corner of the tower of London to the left of the visitor entrance. It is a twin tower design from the sixteenth century in limestone, separate from the white tower and overlooking the buildings where the Royal Mint formerly stood.

Part of the mint's problem, and one Joseph had to face, was the amount of copper coinage in circulation and the due value given it by merchants. At the time of Joseph's joining the mint, copper coinage failed to become legal tender if it was given in values above 6d (six pence). Even by 1757 Joseph laconically remarked that: 'copper coins are not, among us, properly money, but a mere token of exchange'. This problem was partly due to large fluctuations in the value of copper and the flooding of the market with lightweight versions of copper coins formerly minted in various parts of the country. This naturally resulted in the refusal by many merchants to deal in anything other than gold or silver.[8]

In August 1748, in an attempt to research some sort of standard for the production of copper coins, Joseph visited the brass and copper smelting works in Bristol. He was impressed that so much wood and coal was available to the works and especially captivated by the copper smelting process in furnaces using coal preferentially. Whether Joseph was considering such a furnace for use at the mint is open to interpretation but he was fascinated by the abilities of the Newcomen engine, a device primarily used to pump water out of mines, which could have been used as a ventilation engine at the mint, though there is no account of one ever being used there. Problems with copper and its use in coinage did not remedy itself, however, and the situation became so bad that between 1755 and 1779 no copper coin was issued by the mint. In all probability, Joseph's interests in smelting took him to Bristol as an aside from a trip home to Trefecca but it would have given him pause for thought over the use of copper in coins.[9]

The correct standard was required and that was something that would take many years to accomplish when it came to the value of copper, which was a relatively base metal in comparison to gold and silver. It was apparent to him that the value of such coins could only be guaranteed if there were standards that were universally applied, trusted and used nationally. In order to understand and ascribe such a standard, he wrote *An Essay upon Money and Coins* (part I), which demonstrated that coinage had to be tied to economic stability and the sources of wealth in such a way that money became that standard measure due to its inherent trustworthiness. To achieve such trust, coins would have to be minted from metals that had regulated rates of exchange, such as gold and silver, and that were held to be of value by the general populace. Unfortunately, during Joseph's tenure at the Mint, the amount of silver coinage in production declined due to the huge influx of the metal from South America reducing its prices as there was a glut in the market, so gold became the preferred metal of value. By 1758 no silver coin was minted, a situation that lasted for almost thirty years.[10]

It became Joseph's duty to achieve this standard. As a result, he patented an instrument, possibly some form of hydrostatic balance, for the testing of coins that ensured both that each minted coin was monometallic, and that the accuracy of individual weight and adherence was strictly regulated. In this way a guarantee of trust was established in an accurate coinage with a given value. Joseph further reinforced this issue by writing part II of his *Essay upon Money and Coins* in 1758, subtitled: 'wherein is shewed that the established standard of money should not be violated or altered under any pretence whatsoever'. This book was dedicated to the Chancellor of the Exchequer, Henry Bilson-Legge, indicating the high political offices now taking Joseph's advice. Within each chapter Joseph reveals knowledge of general philosophy and a penetrating insight into its political and social application in that he studied John Locke's ideas on 'Monometallism' and applied them to the gold standard.[11]

From 1753 onward, Joseph received a pension of £300 per year from King George II. Although now an occasional parliamentary adviser, in his typical modesty he failed to mention the full extent of his involvement to his friends or family. The dedication of his essay on money and coins to both Arundell and Bilson-Legge above is evidence that, despite his modesty, he was increasingly drawn into the circles of power of his day. Confirmation of his esteem in such circles can be found inscribed on the tablet in St Gwendoline's church: 'His political talents were well known to the ministers of power in his days, who failed not to improve on all the wise and learned ideas which greatness of mind, candour with love of his country led him to communicate'.[12]

In 1757, Joseph experienced a long period of illness which resulted in him accepting as an assistant a deputy named Stanesby Alchorne who was as interested in maintaining the value of coins as was the assay master. Alchorne became assay master in his turn and was commissioned by the government to go to France and report upon their coinage, a task he completed with his essay *Observations on the coins and coinage of France and Flanders collected from the mints of Paris Rouen, Lille and Brussels* in 1777. It is also possible that Alchorne wrote or finalised part III of *An Essay upon Money and Coins* after Joseph's death in 1764. Clearly, they were of a like mind and no doubt enjoyed a good working relationship. Sadly, many of the problems identified by Joseph carried over into Alchorne's time as assay master and some historians record him in an unfavourable light, despite acknowledging Alchorne to have achieved an unsurpassed standard of assaying. Clearly, continued problems with the acceptance of new coins and their value continued to dog the Mint for many years despite the best efforts of its staff.[13]

From this time forward, Joseph was becoming more and more involved with the politicians of his day. As the work of the Mint was answerable to Parliament – though the House of Commons and the Lords were relatively slow in implementing necessary changes to coinage and standards of weights and measures – Joseph became involved in 1758 with the Carysfort commission, which made one of the initial inquiries into the standards across the UK. This initial investigation of the system of weights and measures then in place in industry, coinage and the market shall be dealt with in a later chapter.

Such acknowledgement reflects the diversity of Joseph's interests and grasp of many fields of expertise in addition to the trust other scholars placed in his works. Doubtless Joseph felt a quiet contentment that he had made such substantial scientific and political contributions to the life of his country. His inner modesty forbade him to make more of himself than was necessary and this retiring nature may have resulted in his being passed over for (or more likely refusing) election to the Royal Society despite his close association with the members.

His involvement with the Society continued nevertheless: in 1741 Joseph writes to Martin Folkes, President of the Royal Society, that he is assaying the gold for the Copley Medal for its production based on Tanner's design. Joseph and Folkes would have had much in common as Folkes had travelled throughout Italy in 1733 and wrote a *Dissertation on the Weights and values of Ancient Coins* and the *Table of English Gold Coins from the 18th Year of King Edward III*. Folkes also revised the work with reference to silver English coins in 1745.[14]

It will forever remain a puzzle that Joseph never became a Fellow of the Royal Society. Given the important scientific and political figures that he knew and worked with, it is just not possible that he was overlooked. He must have refused the Fellowship. Seemingly it was sufficient for him to have laboured in these endeavours, leaving posterity to discern what sort of man he was.

6

JOSEPH'S FAMILY AND RELATIONSHIPS

Today, we tend to think of Talgarth in rural Breconshire, mid Wales and London as being far removed from each other and in Joseph's time it probably took almost a week of travelling to get from one to the other. Still, this did not stop the brothers travelling home or back and forth to London to see each other and exchange family news. The Trefecca letters show how close all three brothers were to each other and to their parents: Joseph kept in regular touch with his mother and the growing religious family that grew up around Howell. He was also a regular visitor to his brother Thomas who now lived in London. Being the oldest child probably made Joseph feel a familial responsibility to his younger siblings. Whilst they were growing up, no doubt Joseph took on the role of protector and part-parent and continued to care for and look out for the family's welfare well into adulthood.

Growing up in a rural community in the eighteenth century, one would have assumed that the boys' education would have been rudimentary but, as we have seen, nothing could be further from the truth. The voluminous correspondence between them, their parents and Howell's religious followers tell a different tale. Many of the letters show their erudition, most are in excellent English, some in Welsh and others in Latin. Despite the passage of years, their letters reveal good handwriting, punctuation, attention to detail and an affection for each of the siblings. Joseph always signed his letters off with 'affectionately yours' or 'your affectionate brother' and his brothers did the same. They remained very close despite their distances, religious beliefs and lifestyles.[1]

Joseph and Anne Jones

The letters held at the National Library in Aberystwyth reveal the close bond between the brothers and the measure of trust they placed in each other. The majority of the Trefecca letters began when Joseph moved to London, growing larger in number as Howell started his ministry and with Thomas also moving to London. Across the years they detail their visits to Trefecca, London and the many places visited by Howell, the people they come into contact with, the work they performed, the illnesses they encountered, and the politics and religious life of the country.

The chief interest amongst the early Trefecca letters between 1730 and 1736 lies in the unrequited love that Joseph held for squire Thomas Jones's daughter, Anne. His frequent admonishments to Howell to remember him to Thomas and Anne Jones shows not only the thankful nature of Joseph for the squire's influence on his career but also the extent of Howell's involvement in persuading Anne that Joseph was a respected gentleman, worthy of marriage. Joseph thanks Howell by letter many times between 1732 and 1735 for 'the care he has taken over the affair with Anne' and believes that they could be 'very happy if he [Joseph] could establish himself on a sure footing'. Their letters show how tormented Joseph was over this affair of the heart and his need to impress Anne with stable employment. Over the course of these letters, Joseph had become a teacher of mathematics, had voyaged to the Caribbean under the auspices of the Royal Society and was a respected author and cartographer in his own right. Joseph records sending Anne some books that include *Ancient Geography*, the *Ladies Library* and Jonathan Swift's *Miscellanies*, revealing the breadth of Anne's interests and education.[2]

It may have been that Anne was willing to become engaged to Joseph, but Anne would not marry him until he had become settled in a job that had better longevity than a mere teacher of mathematics. Under her father's guidance and the continued intervention of Howell, she eventually acquiesced and when Joseph joined the Mint (a position that had regular prospects, was universally regarded and was the 'sure

FIGURE 12 St Benet's church outside sign (photograph by author)

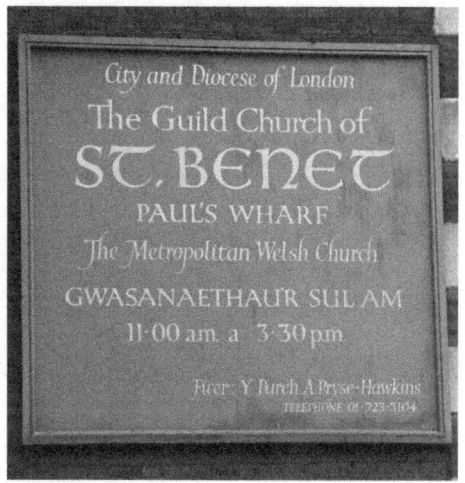

The sign is in gold lettering on a blue background. Close to St Paul's Cathedral, the church itself is set right on the bank of the River Thames and is one of the churches designed by Sir Christopher Wren after the 1666 great fire of London. It is designed in red brick and white cornerstones with a tall tower surmounted by a dome and lantern coated in lead.

footing' that they both desired) Joseph and Anne married on 31 October 1736 at the 'Welsh church', St Benet's at Paul's Wharf, London. Together they would have five children: their only daughter Anna Maria (affectionately termed 'Nanny' by everyone) survived into adulthood and she would return to the Talgarth area. Howell must have been pleased that his efforts on behalf of his brother had materialised, Joseph no doubt being ecstatic that his patience and continued interest in Anne had been rewarded.

Strangely, Joseph's letters are relatively sparse regarding personal details of his family. They include the names of Joseph's daughters but not those of their sons. In seniority they were Mary, Susannah and Anna Maria. Mary was born in 1742 and in one of his letters home Joseph mentions that she is now walking, but no letter records her early death. In one letter of 1742, Joseph chides Howell for not keeping in touch during the illness of one of the boys who subsequently died. In a 1746 letter, Anne gives birth to Susannah, named after Joseph's mother, but again it does not record the date of her death and the letters do not mention the boys or the birth of Anna Maria. Though this seems a curious omission, it is possible that some of the letters may have been mislaid or lost – they are not in the National Library archives. It is tragic that their children died young but, given the unhealthy conditions in

London and the frequent epidemics at the Tower of London, it was inevitable that there would be high infant mortality.

What letters do survive also show the frequency of illness suffered by Joseph and his wife, she after childbirth and he due to various unnamed illnesses, one of which laid him low for six months in 1748 and another in 1757 which necessitated him to get an assistant to help him at the Mint. Joseph was ill for long periods of his life and with the distance of time it is not possible to tease out the causes or names of the various fevers and ailments he suffered. As we have seen, some of them may have been the result of his second Caribbean voyage as he talks of having 'severe seasoning' there, possibly malaria or yellow fever; others may have been due to the toxins at the Mint and unhealthy conditions in the Tower in general.[3]

Joseph and Howell

Nevertheless, Joseph settled into life at the Mint and then became involved with Howell's religious ambitions and tried to influence the course of his adulthood. Although Howell's diary describes how proud he was of his older brother, Howell's life differs very markedly from that of Joseph and youngest brother Thomas.

Howell was obviously a very bright child who grew into an intelligent and forceful man who could have had an excellent life as a rural parson and commentator on local affairs, like so many clergymen of his day such as Theophilus Jones, the historian of Breconshire. Instead, in 1735 he underwent a form of religious conversion during communion at the church at Llangasty Tal y Llyn that set him on the path for which he is now famous, namely the Welsh Methodist revival alongside Daniel Rowland, one who had started his adult life as an Anglican curate before meeting Griffith Jones, a Calvinist preacher who converted to Calvinistic Methodism. Rowland was one of the foremost figures in the Calvinistic Methodist revival then sweeping through Wales. In 1737 Daniel Rowland met Howell and converted him to Calvinistic Methodism. In turn, Howell brought his protégé, the hymn writer William Williams of Pantycelyn, into the Methodist fold.

FIGURE 13
Howell Harris (courtesy and © The History Collection/Alamy)

Howell began to hold meetings in his home and then branched out into travelling and preaching across Wales and then England. He became known as the 'Apostle of Wales', though his Calvinistic Methodism and his style of exhortation no doubt gave rise to the perception of Welsh Methodism as being full of hellfire preachers who concentrated more on sin and damnation than salvation. Nonetheless, despite many setbacks, personal attacks and assaults, Howell, together with Rowland and Williams of Pantycelyn, were to be instrumental in establishing Wales as a haven for religious nonconformity.

After his 1735 epiphany, Howell reports that he has a 'treasure of joy' at finding his calling and, despite Joseph's contacts with Walter Harte at Oxford University, Howell was determined that his form of methodism was the correct path of life and that university was not for him. Joseph was unhappy with this state of affairs. Concerned at the course Howell's life was taking, Joseph travelled to Trefecca in 1735 with the express purpose of persuading his prevaricating brother to matriculate at Oxford University. Howell had been accepted that year for the

FIGURE 14 Trefecca College today (photograph by author)
Front of Trefecca college with a motto surmounting the
door and the front having gabled projections. The college
is painted white and has a central red doorway.

priesthood and training at St Mary's College but was dithering over what to do. Letters between them reveal the unhappy state of Howell's mind, as he was willing to leave the college even before he started and become an itinerant preacher.

Richard Bennett records in *Howell Harris and the Dawn of Revival* that Joseph took a dim view of his brother's preaching activities, a situation that would continue throughout the rest of their lives. His brother's opposition may possibly have been motivated by his association with members of the Royal Society, as Joseph appeared inclined towards a more mechanistic and natural view of the world rather than one that included the hand of God; or, more probably, he was more inclined toward the Anglican general viewpoint. Tom Beynon makes the point in his book *Howell Harris's Visits to London* that Joseph was more probably a Deist, one who believes that God made the world but does not then get involved in any events.[4]

It is more than likely that his distaste for Howell's preaching activities may have arisen due to the popular perception that such 'itinerants' were outside the church – the Anglican denomination refusing to accept them as priests. It is probably just concern for his brother, who may have been stepping away from a lucrative career as a country parson, that led Joseph to think the negative thoughts Bennett attributes to him. He encouraged Howell to stay at Oxford and get his degree in theology but of course this never came to pass. Although the Anglican creed appealed to Howell, his preaching consistently got him into trouble and his request for ordination was turned down three times. Even an opportunity to teach at the Talgarth school was denied him and friends such as Griffith Jones, the priest who had converted Daniel Rowland and an excellent teacher at his circulating schools and Bridget Bevan, the promoter of the circulating schools and the wife of Arthur Bevan, the Judge of equity in north and south Wales advised against ordination due to his young age, though they promised to help him find either a position in the Diocese of Llandaff or a good teaching post if he waited.

Still, by mid 1736 Howell was bound to the path his life would subsequently take and so began the religious arguments he would have with his brothers. In a letter of August 1736 after receiving what he called a 'very severe letter' from his brothers, Howell admonishes Joseph that they 'must not quarrel' and fears that he is troublesome to Joseph. He writes that he hopes Joseph will repent in order to face Christ at the day of judgment and complains that Joseph's letters are too 'intent on affairs of this world'. Although Joseph put up with the style of writing and religious admonishments coming from Howell – indeed, he had written in January 1736 that he 'has no desire to follow Howell's advice (on religious conversion) or to read any books on divinity' – it seems obvious reading between the lines that Howell's brothers were in despair at his passion for methodism. Howell wrote back that he 'believed that Joseph's suspicions regarding his intellect stemmed from the world's opinion of him' and assured his brothers that his intellect was untarnished as, otherwise, how could he be considered to become a teacher?[5]

Howell's intellect was not in question, but his religious commitment was. The trouble of being an itinerant preacher outside the confines of the church affected him deeper in his personal life. It comes as little surprise that Howell was attacked several times by mobs opposed to his methodism but his letters to Joseph uncover a particular disdain for a certain section of society too, which he no doubt targeted. He expressed a bitter resentment that people do not understand religious conviction and complains that for most people 'the drunkard, the dancer and the huntsman is the only man of repute', a dig at both the proletariat and petty bourgeois of his day. Howell did not make it easy for himself and his brothers must have blanched at some of the reports they received about their zealous brother.

Problems also extended into Howell's personal life. He had fallen in love with Anne Williams of Erwood and his letters to her show his earnestness in gaining her as a wife. Anne, however, was totally unsure and was more interested in a local man by the name of John Syms. Although no less a figure than the preacher George Whitefield acted as a go-between, even telling Anne that she should forget Syms, she replied that she was 'adamant that God was her husband more than any man'. Over the course of three years, Howell wooed her by letter and visits despite the opposition of her father and the threats of murder from her brother. She was even locked in the house by her mother in order not to see Howell. Eventually the family relented after an intervention from the landowner and Methodist convert Marmaduke Gwynne and, in 1744, Howell married Anne but it is telling that they were married in the tiny chapel of Ystrad Ffin in Camarthenshire, far from Talgarth or Erwood. Almost cut off from her family and Howell having no affable friends with money, they remained close to the poverty line for many years as they had no real income. However, she remained a devoted wife, accompanying him on his preaching tours and of course looked after the Trefecca *teulu* (family) when Howell was away preaching.[6]

The letters between the brothers show that Howell frequently went to London and stayed with his brothers or with the Methodist preacher George Whitefield. It is also clear that he never gave up admonishing them to accept his religious views. Howell writes that his brothers listened 'with pleasure to his exhortations' despite the fact that the letters between 1736 and 1743 reveal their disagreements and even family tensions. Joseph wrote angrily to Howell in 1742 that he had neglected to ask about Joseph's son, who was very ill at the time, and subsequently died. Apparently, Howell thought that Joseph was far too attached to 'the things of this world, not the next', whilst Joseph totally disagreed with his religious viewpoint especially on things such as dreams and deathbed repentance. Howell riposted with the charge, 'how can the creation appear so beautiful to Joseph and yet he sees no glory in Jesus' face'?[7]

Howell thought that Joseph's activities and achievements merely represented 'head knowledge' but continually encouraged Joseph to reach out to his saviour, Christ. Writing to Joseph in November 1757, Howell laments that 'God forbid that I should find my honest and deep-sighted brother among the anti-Christian unbelievers that would not have the man Christ sovereign over them'. Again, in June 1758 Howell acknowledges that 'in all human enquiries you are my superior' but assures Joseph that this is hardly enough unless he accepts his version of Christ. Howell reports that he is 'pleased that the government takes notice of and rewards Joseph for his continued work' and when reading his publications (*Essay upon Money and Coins*) 'he is filled with pride at being the brother of such a genius' but still wants him to accept his version of Christianity as the only way to salvation. For his part, perhaps in an effort to encourage Howell and to reveal how much his ministry had made inroads into the fabric of religious life in Britain, Joseph reported that he had heard favourable things about Howell from Richard Nugent (1st Earl Nugent) in a conversation with Arthur Onslow, the Speaker of the House of Commons. Nevertheless, the brothers remained at odds in their religious outlook.[8]

It must have been annoying to Joseph, and possibly Thomas also, to have their brother constantly deriding them for having different religious opinions and being nagged to accept a version of nonconformism they could not agree upon. It would appear in the end that the brothers agreed to disagree and put up with Howell's preaching activities, although, even after Joseph's death, his daughter Anna Maria writes to Howell protesting that she is not a sinner and that his version of Christianity was not acceptable to her. Howell simply would not give up.[9]

Despite these problems, the affection they all felt for each other extended to their parents. After the death of their father in 1731 when

Joseph was in the Caribbean, Howell took control of the smallholding at Trefecca Isaf and moved in with his mother to look after her as much as he could. Letters between the family siblings show their affection and care for their mother, with Susannah writing many times to the brothers expressing her contentment and her pride, and in her later years hinting at her age and failing eyesight when in 1739 she asks Howell to 'write in a larger hand' and to bring her a good pair of spectacles when he returns to Trefecca. Once Howell began his preaching activities in earnest and travelled around the country, his wife on occasion looked after her mother-in-law, though at one point in 1750 when their mother Susannah was getting old and weak, both Thomas and Joseph mention that they had been looking after her all summer.

Susannah was very proud of her boys and would have encouraged Howell to continue his activities whilst also acknowledging how busy both Thomas and Joseph were. Howell's preaching must have had an effect on his mother as her letters show her sending pious sentiments to Joseph and Thomas over a period of time. Nevertheless, the letters reveal not just a constant traffic of words back home but the many personal visits made by Joseph and Thomas despite the distance between them.[10]

Where was Howell in 1750 and why did Thomas and Joseph take the time to look after their mother? This date marked the height of Howell's involvement with the so-called 'Sidney Griffiths affair' so it is possible that Howell was laying low and evaluating his future, knowing that his mother would be in good hands. Howell had become convinced that Sidney Griffiths, a young lady from the respected Wynne family, was a prophetess and was 'a divine guide in human flesh' whom he then took on his preaching tours and even to the council meetings of the Welsh Methodist Association, which put him at odds with his fellow Methodists, leading to them abandoning him for the more balanced Daniel Rowland, a split that took ten years to heal. The affair led to universal condemnation from his brothers too, which Howell responded to with his usual bombastic ideals condemning their 'imperfect moralities' and informing them that 'reason is blind without supernatural assistance'.[11]

The situation did not prevent Howell from openly preaching or from leaving the family behind. Filling the gap, it seems that both Joseph and Thomas spent as much time with their mother as they could in her final years. Joseph writes to Howell in February 1751 of their mother's death (Susannah died in January 1751) but how long they had stayed with her after the summer is not known. Susannah reported in a letter of November 1750 to Joseph (who had returned to London) that her health was good so her final illness must have come as a shock to the brothers. In all probability Joseph was with her for some time over the winter, possibly even at her death, as it is he rather than Thomas who writes to Howell to inform him.

Returning to Trefecca after his mother's death and the catastrophic split with Daniel Rowland that threatened to derail methodism in Wales, Howell gathered around him his faithful followers and set up the Trefecca community or 'family' (*teulu*). This family would grow to over one hundred persons, many of whom would rent or purchase auxiliary farms in the area to support themselves and the community. Here, Howell could bring his preaching and lifestyle up to maximum effect, effectively setting up an independent religious commune based upon his own ideology. A few years later in 1758, Howell set up a printing press at Trefecca and one of its greatest accomplishments was printing the Welsh Bible of Peter Williams. It also printed Howell's autobiography and one of the first periodicals in Welsh, *Cylchgrawn Cymraeg*, which included religious and secular articles.

Howell's fame drew the attention of Selina Hastings, the Countess of Huntington who had been drawn to methodism by John Wesley and was a regular attender at the Moorfields Tabernacle chapel of George Whitefield, who eventually became her personal chaplain. The Countess tended toward the more Calvinistic approach of methodism which of course appealed greatly to Howell as it was of a more fundamental nature in keeping with his approach to Christianity. Her influence stretched into parliament of the day and politicians such as Robert Walpole, the first Prime Minister of the UK, Philip Stanhope the 4th earl of Chesterfield and Henry St John MP the 1st Viscount Bolingbroke came to her London home to hear Whitefield preach.

With her large fortune she was able to support and finance the Methodist movement.

Howell's involvement with her culminated in 1768 with her financing the building of Trefecca college as a place for the training and ordination of men for the purpose of the ministry. George Whitefield preached at the opening of the college, which grew so large that in 1792 it was moved to Hertfordshire and eventually Cambridge where it merged with Westminster college, the training school for Presbyterian Church of England. Today Trefecca college is still at its original site in Wales, a testament to her and to Howell's foresight, and is still used as a training college for ministers in addition to housing a museum dedicated to Howell's work.[12]

Despite the fame of his religious brother, Thomas remained in the background with his head down. Like Joseph, he was disinclined to accept Howell's version of Christianity but perhaps unlike Joseph he was not prepared to dispute with him. In reality, all three brothers had more pressing engagements to contend with: Howell with his itinerant activities and Joseph and Thomas at the Mint and in the tailor's shop.

The differences between the brothers in religious outlook did not sour their filial relationship and they remained devoted to one another over the years. Indeed, Joseph came to increasingly depend on Howell at Trefecca as the Tredustan estates of his father-in-law Thomas Jones had passed down to him through his wife Anne. Several letters from Joseph to Howell are concerned with repairs to Tredustan and management of some of the homes and farms on the estate. Living permanently in London, the house had to be looked after and an annual rent was agreed with Walter Prosser for £55 per year to inhabit the house and manage the estates. Prosser must have been a prosperous landowner in the area as he already had stock in Trefecca uchaf farm. Howell lamented to Joseph that he was unhappy that he may lose Trefecca uchaf and was not on good terms with Prosser, hoping that Joseph would actually take up residence at Tredustan or allow Howell to do so. Although from their letters it seems that Joseph wanted to move back to Trefecca at some point, he regretfully wrote to Howell in April 1757 that this was no longer possible. This situation dragged on with no resolution right up until Joseph's death when his daughter Anna Maria inherited the estates and returned to the area.[13]

However, their stand on Howell's religious outlook did not prevent them from taking a keen interest in the development of the Trefecca community under their brother's leadership. Davies records in his essay *Trevecka (1706–1964)* that Joseph gave freely of his ideas in the building and administration of the sixty or more trades carried on there. Joseph also encouraged the use of new farming machinery such as seed drills, ploughs and horse hoes which were coming into service. Joseph was also instrumental in assisting Howell integrate with the Brecknockshire Agricultural Society.

Founded in March 1755, it became the first such society in Britain. Sixteen gentlemen of the area met at the Golden Lion inn with a view to setting up a hunting party but this soon evolved into a society that promoted agriculture and husbandry. Initially started by Charles Powell of Castell Madoc near Lower Chapel, he was joined by Colonel Williams of Gwernyfed and Pendry Williams of Penpont. They advertised for additional members in the *Gentlemen's Magazine* and resolved to 'do something to support and encourage agriculture, manufacturing and promote the general good of the county of Brecon'. The society has met without a single break since on the Watton fields at Brecon and is now a limited company with charitable status. What is even more surprising is that all its staff are volunteers.[14]

This was an important contribution as Britain was undergoing an agrarian revolution that enabled a large increase in crop yield by instituting crop rotation, selective livestock breeding and encouraging the enclosure of lands, which was to provide a new pattern of farming for the next two centuries. Under Howell's influence, the society turned from a mere drinking and meeting club for gentlemen into a dynamic and forward-looking organisation that assisted many of the surrounding farmers and charitably gave the poor work to undertake.

Utilising some of these methods, the Trefecca family produced ample foodstuffs and any excess was sold at market. The community was renowned in the locality as an industrious wool producer, and Joseph and Howell were instrumental in setting up Brecon market in 1756 as an outlet for such activities and to cultivate a market for quality products across the county. Thus, Joseph played a small part in the agricultural processes then assisting the burgeoning Industrial Revolution in which Wales had a central role.[15]

In 1763, the Trefecca community experienced a highlight that was of pivotal importance to Methodism in Wales: the visit of John Wesley, the founder of the Methodist movement. Joseph was probably not in attendance at Trefecca on this occasion as not only was he extremely ill as he reported by letter to Howell in March, but also mourning Anne, his wife of twenty-five years who had died in April of that year, leaving their daughter Anna Maria to look after him. He was also, when he could get to work, increasingly busy at the Mint, meeting the French polymath and astronomer Jérôme Lalande on 6th April that year. Lalande recorded a snippet of their conversation in his diary, with Harris promising him a standard weight when one was available, which revealed how concerned the French were with standardisation of their coinage.

The death of his wife deeply affected Joseph and he remained in ill health until his own death the following year. Still, Joseph and Thomas remained close during this time and the letters at the National Library record Thomas writing to Howell informing him that 'Nanny' (Anna Maria) had paid them a visit without Joseph, revealing that she was now an independent young woman though still devoted to her father.

The Jacobite Rebellions and the Trefecca Militia

Against this background, the political state of the country was in upheaval. During the lifetime of the brothers, the Jacobite rebellion came to a head in 1745 with an invasion of England from Scotland by Charles Stuart otherwise known as 'Bonnie Prince Charlie' in order to secure the throne of England for his father James Stuart. This situation had arisen as there was almost no legitimate monarch (as seen by the Jacobites) in Britain apart from those who could be brought in from abroad under the 1701 Act of Settlement.

The problem began at the 'Glorious Revolution' in 1688 which replaced James II, a Catholic monarch, with William and Mary who were ardent Protestants. When Mary died in 1694, and William in 1702, her sister Anne took over as Mary had no children as inheritors. Similarly, when Anne died leaving no children despite having seventeen pregnancies, the only eligible one left was their Catholic half-brother James Francis Edward or Anne's heir apparent, Sophie the Electress of Hanover. Unfortunately, Sophie had died two months before Anne and this pushed the monarchy and the constitution toward its distant relations in Europe. The Act of Settlement stipulated that the English crown could only be passed on to a Protestant king or queen, not the Catholic James Edward, which left the only candidate as Sophie's son Georg Ludwig of Hanover who subsequently became George I.

Uprisings by the Scots supporting a Catholic king were suppressed in 1689 and again in 1708, 1715 and 1719. The most successful of the uprisings was of course that of Bonnie Prince Charlie in 1745 who, upon capturing Edinburgh, then marched south toward London. However, lack of support from Welsh and English Jacobites and no simultaneous invasion from French sympathisers put the rebellion into doubt and the Jacobites turned back at Derby. The decision to retreat caused a split between Charles and his supporters in Scotland and his defeat at the battle of Culloden in April 1746 sealed the Jacobites' fate. Charles escaped to France but could never regain support and he died in Rome in 1788, thus ending all hopes for a Scottish and Catholic monarchic restoration.

This internal British strife took place against the background of the War of the Austrian Succession in which the armies of Britain played a major part. The Jacobites took advantage of the fact that most of the British army was abroad fighting and hoped to capture London and place their king on the throne. The situation frightened many people, including Joseph's wife Anne who wrote to Howell in January 1746 about her fear of the soldiers coming to London and what would happen to the country under catholic rule. Their support for the monarchy under King George II was already evident and the thought of the chaos and potential civil war that would ensue under Bonnie Prince Charlie must have been extremely troubling. This was especially so as Joseph was extremely busy with the transfer and minting of gold and silver from the captured French ships and so Anne, being left alone and feeling ill much of the time after childbirth, was naturally fearful.[16]

The uncertain political nature of the country and its fractious relationships with countries on the continent were to result in the 1756 outbreak of the Seven Years' War against France. Six members of the Trefecca *teulu* (family) immediately joined the militia at Hereford and went to fight with James Wolfe in Quebec. Wolfe himself was killed in the battle for the Plains of Abraham and four of the Trefecca volunteers were also killed in this battle and two wounded, one with a 'bayonet through the tongue' as Howell recounted in a letter to Joseph in January 1760. This did not deter the volunteers of the Trefecca community; as encouraged by Joseph, Howell himself became a commissioned officer in the Breconshire Militia and was joined by twenty-four of the able-bodied youths of the family, which became known informally as the Trefecca Militia. Uniforms for the militia were provided by Thomas, who wrote to Howell enclosing his bill and wishing him and the militia every success. Howell wrote that the soldiers from Trefecca are 'of no military spirit, but are armed with the true principle of loyalty to our God, King and love for our country'.[17]

In the event, Howell was never called upon to fight but he enjoyed his rank of Captain and used the travelling opportunities it presented to further preach his form of the gospel in such places as Bideford and Norwich, even taking some of the militia to protect him, as was the case in his discourse at Great Yarmouth where the sight of armed men surrounding Howell put an end to the crowd's initial efforts to bring him to harm. Howell complained in many letters to his brothers and his followers that a military life was far from what he wished to do but he was compelled by duty to God, country and King in order to vanquish what he called the 'tyrannical spirit of popery'.[18]

At the end of the war, the soldiers returned to Trefecca with their arms and Howell's musket and sword are on display in the small museum in the theological college there. The Trefecca militia became subsumed in the larger Brecon Militia but the letters from officers in the Trefecca archives give a glowing testimony to the thankfulness of

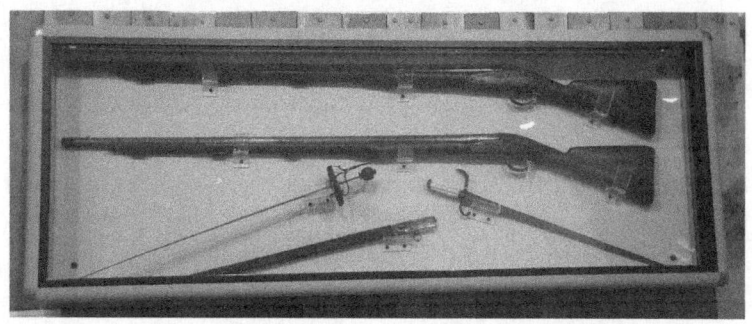

FIGURE 15 Howell's musket and swords (photograph by author)
A glass case containing two long musket rifles and two swords. One is a ceremonial officer's sword with scabbard and the other is a rapier without a scabbard.

the army at a time when such volunteers were most needed. Howell's foray into the militia was eventually curtailed when he was contacted by Daniel Rowland and William Williams to return to the fold of Methodism and return to 'fill up your place amongst us, which we sensibly acknowledge has been long vacant'. Howell of course was eager to oblige and continue his ministry and the militia was eventually disbanded after Howell visited Joseph at the Mint in October 1762 with the promise that troops from Germany would come and take over the duties of the soldiers.[19]

Sadly, whilst Howell was away with the militia, a smallpox outbreak destroyed the lives of eighty of the Trefecca family and it was to a smaller and devastated *teulu* that he would return. In 1764, the family were recovering when an old acquaintance of Howell's returned to Talgarth. Selina, Countess of Huntingdon had accompanied Howell on his preaching activities across Wales and they had been introduced to each other by George Whitefield as early as 1748. She had come to Howell with the idea of establishing a training college to prepare men for the Calvinistic Methodist ministry and to do so she leased Trefecca Isaf, the Harris's old home, from brother Thomas who now possessed it after his mother's death. Upon the site, they built a college which is now known as the college farm. There Howell would preach to the students twice a week and, upon its opening, Howell's mentor and friend

George Whitefield addressed a crowd of thousands who had come to see the college.

Howell died in 1773 at the age of fifty-nine. Records reveal that he had been ill for some time and was worn out by his preaching and the stresses of looking after the Trefecca family. David Ceri Jones, historian at Aberystwyth University, concedes that Howell was a deeply flawed leader, authoritarian, arrogant and belligerent and his relationship with Sidney Griffiths brought almost the entire methodism movement into disrepute. Nevertheless, the love that many bore for him was evident at his funeral which was attended by over 20,000 people. His body rests near the altar in St Gwendoline's church in Talgarth close to his memorial tablet. Of the three brothers it seems that only Joseph is not buried in the country of his birth.[20]

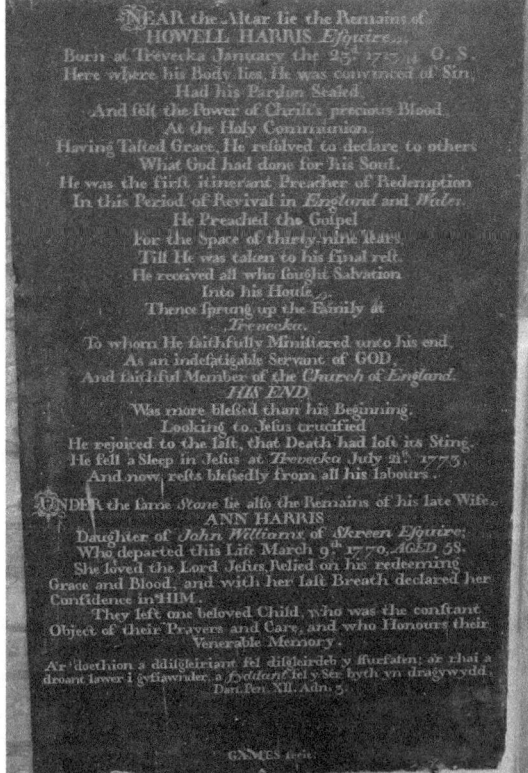

FIGURE 16
Howell's memorial tablet (photograph by author)

The memorial tablet of Howell Harris. It is rectangular slate with inlaid gold lettering recounting his ministry and a short description of his life.

The Fate of Thomas Harris

The close friendship and family ties that bound the brothers endured throughout their lives. We have seen how Joseph and Thomas made visits to Trefecca as frequently as they could, and Howell occasionally visited them both in London during his visits to preach at the Moorfields Tabernacle established by George Whitefield. The close relationships they shared as children were maintained throughout their lives despite their religious and secular differences. Joseph's life in London had continued in the typical vein of many scientifically inclined men of his time. Until his death in 1764, he would occasionally attend Royal Society meetings and catch up with Howell on his regular visits to the capital to preach, or by calls upon his brother Thomas and his partner and children, who had been living in London for many years as a prosperous tailor with a shop in the Strand.

Thomas, born in 1705, was the youngest of the three brothers and seems to have had a chequered history in comparison to his siblings. His partner's name is not recorded, nor is the date or location where she died, possibly because she was a common law wife and Thomas was not married to her. In 1728 he was sent by the family to the town of Bath where he began to learn the trade of tailoring. From there he moved to London to work for his uncle, the tailor Solomon Price. Thomas soon set up on his own but his business suffered as a result of a profligate lifestyle and he was forced to go to France for a while, possibly to escape creditors. Both Howell and Joseph wrote to him in April 1736 trying to dissuade him of this plan but it seems that Thomas was adamant in going. Howell sends Thomas a letter in September 1737, pleased that he has arrived in Paris and sends the affections of their mother who hopes he will return to his native country as soon as possible. In the event, Thomas returned to London in January 1738 and was at first employed as a tailor in the Strand, eventually again establishing himself in independent business. Thomas finally made a major business breakthrough when he managed to secure army uniform contracts.[21]

Throughout the eighteenth century, Britain was engaged in a number of conflicts and there was a constant demand for uniforms. The famous Red Coats worn by British soldiers had evolved throughout the English civil war, being the uniform for Cromwell's New Model Army, but by the beginning of the eighteenth century, the Colonels of each regiment were responsible for acquiring uniforms for their men. This changed in 1707 when a royal warrant established a board of generals to regulate the clothing of the Army and any uniform then supplied had to conform to patterns approved by the board. Red dye made from cochineal or madder (a plant root) had been in use for almost a century due to its cheapness and the fact that it withstood the weather without fading. Uniforms were therefore to be mass produced and to be as cheap as possible but even then were primarily worn by officers and sergeants rather than the private soldier.

The army contracts were a boon to Thomas's business as he now had a customer who would demand a constant flow of materials. This enabled Thomas to amass a fortune which he later used to acquire the Trefecca Fawr and Tregunter estates. Retiring to the Brecon Area in 1768, he built a large house at Tregunter, which was completed in 1770, and he became a landed country gentleman who went on to serve as Sheriff of Breconshire and head of the Brecon Militia. Only some of the foundations of the Tregunter house remain as it was demolished in 1925. Its ornate iron railings were removed and now adorn the front of Talgarth Town Hall.

Thomas had three children of whom not much is recorded even in the Trefecca letters. His daughter Elizabeth (Betty) Robinson lived with him at Tregunter and was a committed Methodist, being a regular attender at services at Lady Huntingdon's college at Trefecca which, of course, her uncle Howell had established. It is unknown if she married or had children or lived in a house on the estate as the last family member to actually own the house was Eliza Anne, Joseph's granddaughter, which seems to suggest that she died without any surviving children.

Thomas had a son, who was named Thomas Robinson, possibly a reference to his mother's surname. Very little is known about his life apart from a few salient facts due to his associations. Apparently, he was a drunk and a troublemaker who nonetheless appeared to have good social connections, probably thanks to his father. He was apprenticed in London to a merchant company as an articled clerk and, through his connections, went on to marry a famous actress, Mary Darby known as 'Perdita' in 1773. Although she took his name and was known as Mary Robinson, it was not exactly a match made in heaven; the marriage was doomed due to his drinking and gambling and her affairs. Thomas Robinson fell into debt in 1774 and he and his wife went to live with his father at Tregunter. Here, Mary gave birth to their daughter Maria Elizabeth Robinson. However, they were soon arrested by their creditors for their debts and sent to Fleet Prison in 1775. It seems that Thomas Harris was not going to help out his recalcitrant son or the mother of his granddaughter.[22]

Whilst in prison Thomas Robinson refused to work to pay off his creditors, so Mary took up writing and looking after their little daughter who had gone into prison with them. Mary could not bear to be parted from her and would not leave the child with any relatives. Whilst there, Mary wrote several poetical works which appear to have been not well received until they attracted the attention of Georgiana Cavendish, the Duchess of Devonshire, who supported her and became a patron. The next year, upon their release, she became an actress with Thomas's concession and was engaged by the playwright Sheridan.

Mary eventually came to the attention of the Prince of Wales (George IV) and they had a brief affair after he paid off the bond of £20,000 on her acting career. This affair led to her estrangement and divorce from Thomas Robinson. She later blackmailed the prince over his love letters and received a payment of £5,000 and an annuity. Mary became so famous in acting and literary circles that in 1781 she had her portrait painted by none other than the artist Thomas Gainsborough and another a year later by Joshua Reynolds. She moved around the continent, briefly staying in France and Germany before arriving back home in London where she became acquainted with Mary Woolstonecraft and William Godwin and the poet Samuel Taylor Coleridge. Toward the end of her life at the age of forty-four, she moved in with her daughter Maria at Englefield Green near Egham where she died in December 1800. Thomas Robinson was granted the rights to her estate and in 1803 he acquired the estate of his half-brother William, of whom little is known.[23]

Maria went on to become a writer and editor of her mother's poetic works and autobiography. In 1794 she published a novel *The Shrine of Bertha* and in 1804 she published an anthology of poetry called *The Wild Wreath* dedicated to Frederica Charlotte the Duchess of York, which suggests that she had some acquaintances in the realms of power and influence.

Of the third child of Thomas Harris, history is silent, possibly due to the fact that the child may have died before reaching adulthood. Thomas Harris died in 1782 and, probably unsurprisingly, ignoring his own children, left his estates to his niece Anna Maria, the daughter of Joseph Harris.

The brothers all seem remarkable in that initially one would have thought them doomed to a rural servitude just as many of their contemporaries were. Blacksmith, parson and tailor, they would have been lost in time if it were not for their incredible drive and determination coupled to a precocious brilliance that brought them to the attention of the gentry and people who could influence their paths and careers. Together, the Harris brothers are a testament to personality, drive, hard work, education and application that brought them to national fame. Joseph's heirs and legacy, we will deal with in another chapter.

7

THE CARYSFORT COMMISSION

Joseph had been Assay Master at the Mint for almost ten years before he was drawn into an engaging and far-reaching commission from parliamentarians. The Huntingdon MP John Proby, 1st Baron Carysfort was returning to Parliament in 1757 after working as an admiral of the Royal Navy and was concerned with the prices, measures and weights of various agricultural products, but especially corn. He contacted Joseph and commissioned him to make some standard weights against which things could be measured fairly. Joseph subsequently constructed a mahogany box and ten grain weights for Carysfort's 1758–9 Select Committee on weights and measures which was to 'inquire into the original standards and measures in this kingdom, and to consider the laws relating thereto'.[1]

It appears that personal rather than governmental interest lay behind Carysfort's attempts to standardise weights and measures. He may have acted when and why he did because in the previous year a Commons select committee investigating the high price of corn had complained that the uncertainty of the available measures encouraged fraud and that many problems could be overcome by selling corn by weight rather than volume. His interest in this respect led him to one of the few men who could understand the problem of accuracy in measurement and standardisation. Joseph Harris was already well known to MPs and had direct access to them if he needed it via Richard Arundel MP who was the Master of the Mint, Lord Commissioner of the Treasury and, of course, a Fellow of the Royal Society.

No real examination of weights and measures had been made since the time of Henry III in 1266 and, although updated, changed and reordered by various sovereigns since then, there was still no universally accepted standard for any measurement of any commodity in the UK. The disorganised and arbitrary way in which things were measured and valued can be judged from this:

> an English penny, called the Sterling round, without clipping should weigh thirty two grains of wheat, taken from the middle of the ear and twenty pennies to make an ounce. Twelve ounces would make a pound and eight pounds a gallon of wine and eight gallons of wine a bushel.

Today such measures sound strange to the ear and we can immediately take issue with twelve ounces to the pound. In addition, most commodities were also sold by their volume and such volumes would differ between dry goods and liquids due to density. The traditional corn gallon measured 268.8 cubic inches but an ale gallon measured 282 cubic inches whilst a wine gallon measured only 231 cubic inches.[2]

Clearly there was a lot to be done! Standards had to be placed on accurate, measurable and acceptable premises, not the weight of grains of wheat, or the number of cubic inches in a gallon, which could vary across a range of quantities and dimensions. Furthermore, the units used in Wales, Scotland, France and England all differed markedly. Considering that the Act of Union in 1707 married Scotland to England, a union of measurements was sorely needed. Additionally, imports of wine and other goods from France, which differed from those of the UK, meant that a true and standard series of weights and measures had to be found. Just as an example, in England the (Imperial) gallon equalled 4 quarts, 8 pints or 32 gills and contained 231 cubic inches of fluid or the equivalent of 3.78 litres of French wine, the Scottish gallon equalled 3 imperial gallons and the French did not even use the gallon and their measurement of litres changed over several decades.[3]

Additionally, there were three separate standards of measurement, many of which differed depending on geography. First, there is the Avoirdupois system which dealt with weights in ounces, pounds, stones, hundredweights and tons. Secondly, there was the Troy system of weights dealing with precious metals and gems in which a Troy pound was actually lighter than an Avoirdupois pound. Lastly, there was the Apothecaries system which was similar to the Troy system but used different units for lighter weights such as the grain and in which there were twelve ounces to the pound. This system was not abolished until as late as 1978!

Into this arena stepped Joseph who, as Assay Master, was in a good position to advise on how and why such standards were necessary and could be applied. Joseph was not a member of the parliamentary committee as he was not an MP, he merely advised as a consultant and lent his expertise when necessary. In the following year, 1758, the Commons received a select committee report into weights and measures from the commission and a year later, in 1759, a second report followed, again from Carysfort. These committees, especially the first, formed the basis of all subsequent discussion of the problems. They were principally concerned with the inadequacies of existing legislation and enforcement but sadly they largely ignored the issue of just how varied common usage of weights and measures were. Much of the report of the first committee was given over to an outline of the numerous statutes, their limitations and contradictions, the physical measures that existed at the Exchequer and the difficulty of regulating the production of new weights and measures.

No doubt Joseph, as consultant to the commission, would have highlighted these details and put forward suggestions as to what changes would be required. How much of his advice was heeded can be seen from the report; though he did not personally record his contributions, the committee fairly represented the many commissions he was tasked with. Under the committee, the whole chaos of this unregulated system was brought to the attention of Parliament, and a series of resolutions amounted to one of the first real efforts in over 150 years to standardise the weights and measures system.

Joseph was consulted frequently over a period of several weeks and the report states that he was asked his opinion on measures and capacities then in use. His response was that 'the gallon is our old standard for the bushel and hogshead … but of the gallon itself, there was no real standard because, with liquors, they could not be gauged or measured exactly. What is necessary is to specify the number of cubic inches.' He went on to advise the committee that without a proper standard measure, 'suspicion arises despite the lack of consequences as to differing measurements to the existing 271 cubic inches'. Joseph obviously thought that rounding up was both a fairer and easier way of measuring the gallon and standardising it at a point where all liquids and solids such as corn grains that filled the said vessel would then agree. Two of the committee immediately agreed with Joseph and his proposal was subsequently accepted.[4]

Thereafter, the committee requested the instrument maker and artisan John Bird, under Joseph's direction, to make a gallon vessel as the absolute standard that would be placed in the exchequer as the UK arbiter of measures of the gallon. This vessel was to contain 300 cubic inches, exactly as Joseph had proposed. Joseph would have undoubtedly known Bird as they were both protégés of sorts to John Sissons, whose shop in the Strand was well known to Joseph during his apprentice years.

As all measurements were under scrutiny, Joseph was also asked his opinion of the standard yard and rod. The imperial yard as we all know today is standardised at thirty-six inches, but it is interesting to see that Joseph initially recommended its length to be

> 38 or 39 inches long and a stick should be made constructed ... that is 39 inches long and an inch thick. This rod could then be marked with a standard yard of 36 inches should then be placed in safe custody and used occasionally as the sizing of yards and kept within the Exchequer.

Joseph felt that the existing measuring rods were too small and they should be updated. It seems incredible today that Joseph's determination of the standard rod length for the yard is so close to the standard metric metre! Why did Joseph pick this particular size? Possibly because another measurement of length was also in use in Britain. Called the Ell, the size of one Ell was 45 inches but of course varied in length between England, Wales and Scotland. Perhaps Joseph was attempting to bring in a single measurement that was simpler than trying to understand many different units. Once again, the committee accepted the proposition; John Bird made the standard rod and it was placed in the keeping of the Government.[5]

Sadly, Joseph's one-pound standard Troy weight was destroyed in the 1838 fire of Westminster Palace where the imperial standards were kept. John Bird's rods survived and are still in the House of Commons today. New standard yards were later constructed by the Royal Astronomical Society and they followed Joseph's tradition but later were found to be shrinking at the rate of one part per million every twenty years. Thereafter, as comparisons with the standard metre were made in 1895, and the metre found to be true, the adoption of the yard as 36 inches in relation to the metre's 39.370 inches was made. The new standard yard became today's representative measure and was adopted by Parliament and across the Commonwealth (and also in the USA) as the new International Yard in 1959. Joseph's work had come a long way. However, his consultancy went further.

The pound (lb) also had to be standardised. Joseph suggested that the Troy pound be adopted as standard weight as it had been in use the longest and it also was used in the weighing and measuring of coinage, an area he was well suited to comment upon. The weights used at the Mint were of course Troy weights and were made to a very accurate standard, far better than the existing Avoirdupois pound. Joseph stated that the 'troy weight was easier to divide into smaller units and was in use across the world' and therefore was a more accurate measure of weight agreed upon by various nations. To ascertain the absolute standard, they used Joseph's 'curious and exact' pair of scales that he used at the Mint to weigh gold, finding that the Troy pound was in fact heavier than the Avoirdupois pound. The committee then commissioned Joseph to make three exact Troy pound weights that could be used as the standard measure, which of course he went on to construct.[6]

Upon the conclusion of the committee's dealings, the board resolved that the standards that would now go forward for ratification by parliament should all follow Joseph's recommendations with the exception of the gallon, which was to remain at 282 cubic inches. His determination of the yard, the pound, the Troy pound and measure used in coinage were to be adopted as standard and all other standards previously in use now to be repealed. They also suggested that it would be illegal to use any other measures and that strict penalties resulting in imprisonment be passed into legislature. By the time of the 1760 reading of the committee's resolutions, they were passed by the House of Commons with the proviso that a bill would be produced that would pass the recommendations into law at some future date and that the Carysfort report be the basis of that bill.

Surprisingly, the entire commission was the result of an initial request by a single MP, almost as if somebody had to galvanise the House of Commons into acting on the systems then in use. Considering the importance of standards across the UK on weights and measures it could be assumed that Carysfort's committees were set up by the Government. However, it seems that there is virtually no evidence for this. Carysfort himself was an ordinary Member of Parliament, sitting in the House of Commons from 1747 to 1764, being elected twice and holding only minor office and featuring hardly at all in Thomas Pelham-Holles, the Duke of Newcastle and First Lord of the Treasury's correspondence during this period. Fortunately, the Duke prevailed upon Parliament to meet the expenses of the Carysfort committee, perhaps recognising that there was a need for such a parliamentary inquiry.

This in itself was unusual because, although Holles had become Prime Minister for the second time in 1757, he was a member of the House of Lords, not the Commons. It was probably his influence that allowed the exchequer to pay for the commission, though, despite all this authority, things did not greatly improve on standards until the next century under the Weights and Measures Act of 1824, when many of Joseph's recommendations were adopted. Nevertheless, at least a foundation had been laid, and Joseph had played a large part in this advance despite any personal frustrations he may have felt after the reports were submitted.

Apart from the commissions he chaired, which investigated the problems with the weights and measures system, nothing further suggests that Carysfort was a central figure in policy making. How close a relationship Joseph had with the committee members, who are not even mentioned in the final report, is open to suggestion as Joseph does not record his interactions with them and the report records the work he did for them but little else. However, in *An Essay upon Money and Coins*, he dedicates that work to the Chancellor of the Exchequer Henry Bilson-Legge who would become part of the review team that would oversee the report, so at least he knew one of them relatively well. Bilson-Legge was also a Fellow of the Royal Society, so it is possible that Joseph knew him outside his parliamentary dealings.

There is nothing in the Carysfort report to make the reader consider that Joseph and Carysfort actually had a close working relationship. Joseph merely appears to have acted as a consultant alongside his job as assay master, as he was the only person with the recognised authority to declare on weights and measures in a professional capacity. Additionally, the reports of 1758 and 1759 were not immediately followed up and in 1760 their resolutions were repeated and bills were ordered to be prepared by a committee of seventeen whose membership included Carysfort, the Chancellor of the Exchequer as seen above, the Lord Register of Scotland James Douglas, and the Earl of Morton amongst others, but not Joseph himself.[7]

Unfortunately, the reports were prepared too late in the parliamentary session and the two bills which could have standardised the system both failed to go through Parliament. Nothing was heard of the issue for five years, by which time Joseph had died, when, again, two more bills were introduced, but again so late that they too failed to pass parliamentary scrutiny.

Going a little further and speculating on the relationships between the committee members, it was known that Carysfort was a dilettante who got into debt and had his furniture repossessed by his butcher to whom he owed £219. His family record that he spent £10,000 on a mistress and most of his fortune on 'loose women'. By 1767, although past Joseph's lifetime, Carysfort was out of public life and living in France, dying in 1772 in Lille. Perhaps many MPs had a personal distaste for Carysfort's known lifestyle resulting in delays to the report? We shall probably never know.[8]

The fact that Joseph played a large part in the committee's consultancy is testament to his stature as Assay Master and a recognised authority on accuracy, even inventing a perfect scale for the measurement of gold at the Mint. Such a set of scales and weights would have to be made to exacting proportions so that the weight of gold in the scale was as specific as possible. Several such scales already existed but Joseph's perfectionist approach ensured that standards and accuracy were maintained and improved upon. He was one of the few men that any commission could call upon for advice in such an exacting and far-reaching act of legislation that would survive right into the twenty-first century.

8

THE TRANSIT OF VENUS

Joseph travelled home as regularly as possible to see his brother and to encourage Howell in his agricultural pursuits. Although Joseph himself was probably not particularly religious, he found that he could lend an occasional helping hand advising on the trades within the growing religious family at Trefecca and attending some of the early meetings of the Brecknockshire Agricultural Society, which he promoted. Although approaching sixty, Joseph's health was declining; he was suffering from an unnamed recurring illness but passing it off as gout to his siblings. Nonetheless, he still endeavoured to be an active scientific observer and maintained his role as Assay Master at the Mint. An opportunity to become involved in a wonderful celestial spectacle presented itself in 1761.

One of the most important scientific events of the eighteenth century was the transit of Venus across the Sun, which was predicted to occur on 6th June that year. This was the first of two transits during the eighteenth century. Joseph's friend Edmond Halley had hinted as early as 1691 that good observations of such an event from known longitudes would give an accurate measurement of the Earth–Sun distance. This distance became known as the Astronomical Unit (AU) and was necessary for improved astronomical calculations of the size of the Solar System. Even before Joseph arrived in London, Halley called for the Royal Society in 1716 to coordinate a far-reaching and well-planned international endeavour to observe the 1761 and 1769 transit events. Joseph no doubt recalled the advice of his old mentor and decided to observe the 1761 transit despite the fact that he was not a part of any recognised expedition. He may have been encouraged in doing this by the Earl of Macclesfield who was the serving President of the Royal Society at the time, which hints at Joseph's continued association with this institution.[1]

Transits of the planet Venus across the disc of the Sun are extremely rare celestial occurrences that repeat in a 243-year pattern with two 'eight-year pairs': a pair of transits eight years apart in December and then a gap of 125.5 years followed by the next eight-year pair in June followed by a gap of 105.5 years. The last transits of Venus were in 2008 and 2012, and the phenomena will not be seen again until 2117. Due to its rarity but easy visibility on the solar disc, it was the best of the inferior planets to try and determine the Earth–Sun distance. Although transits of the planet Mercury were more frequent, occurring every seven years, the tiny disc it portended made it very difficult to accurately measure its timings on entry and exit across the solar sphere.

FIGURE 17 Transit of Venus (photograph by author)

The sun in yellow with a black dot traversing it. The black dot is the planet Venus. Photographed by the author during the 2004 Transit of Venus.

Additionally, the problem of solving longitude was absolutely essential to the task as the observations and timings would only make sense if the observers worked at known longitudes, and therefore distances, from a central meridian such as Greenwich.

Although many ancient cultures made reference to the planet Venus, often ascribing its appearance to some god or other, techniques and observation in ancient times were not refined enough to see or to predict transits of the planet. The Mayan civilisation of central America in the second millennium CE were great Venus watchers and mathematically predicted its synodic appearances, a timescale of nineteenth months when Venus appears in the same position in the sky relative to the Sun and Earth. However, we have no evidence that the Mayans ever saw or predicted a transit. It is a great pity that these great Mesoamerican civilisations were unknown to the West when Joseph visited Vera Cruz in 1725 as the great cities such as El Tajin and Cerro de las Mesas would have astonished and delighted him.

The first person to witness a transit event may have been the Persian polymath Ibn Sina known to us as Avicenna, as there was a transit on 10 May 1032 to which he makes reference in his writings. How he observed the transit is not clear but modern scholars think he may have seen a large sunspot instead. The first person to accurately predict a transit of Venus was the great mathematician and scholar Johannes Kepler who in 1627 predicted that the Earth–Sun distance may be obtained trigonometrically from watching a transit. He predicted the 1631 Venusian transit and encouraged many viewers to observe it. However, he was unable to accurately predict the path of the transit and unfortunately it was not visible from Europe.[2]

The next transit in this eight-year pair series was in 1639 and it was on this occasion visible across Europe. The British astronomers Jeremiah Horrocks and William Crabtree both noted the transit on 4 December 1639 but neither saw the entire event and their positions in Preston and Manchester respectively did not give a large enough baseline to gauge the Astronomical Unit, though Horrocks determined from his observations and calculations that the Astronomical Unit was 59.4 million miles, almost 65 per cent of its true value.

The next pair of transits did not occur until 1761 and 1769 and so, with the increased accuracy of observation coupled to advances in instrumentation, the transits of the eighteenth century held great promise in finally determining the Earth–Sun distance. However, as seen above, there were a few things that needed to be accurately known. One such was precise global positioning by means of longitude and the other was the precise timings of the event from locations whose longitude was well known. Subsequently there would be laborious calculations that would take into account these timings from various global positions to ensure accuracy. Although in years gone by this may have resulted in solo efforts by some astronomers, the calculation of the Astronomical Unit was going to take a huge, combined approach.

Subsequently, the 1761 transit became the focus of the first great international scientific effort. History records that the transit was sighted by 121 observers from sixty-six locations, in places as diverse as Newfoundland and Calcutta. British observers such as Charles Mason and Jeremiah Dixon observed the transit from South Africa, whilst John Winthrop went to St John's, Nova Scotia in Canada, Nevil Maskelyne studied it on the island of St Helena and Nathaniel Bliss observed it from Greenwich. Much of the money spent by the Royal Society to partake in this international effort came from Thomas Pelham-Holles, the Duke of Newcastle and Prime Minister of Great Britain for two terms, which reveals the importance to British interests influencing navigation, trade and imperial expansion. The telescope maker and Society fellow James Short collated these disparate observations and computed the results for the Society.

From the above, it would seem that these scientific expeditions were carried out in excellent circumstances without a great deal of fuss. Nothing could be further from the truth as evidenced by some of the French expeditions. Under the guidance of the wonderfully named Guillaume Joseph Hyacinthe Jean-Baptiste Le Gentil de la Galaisière (shortened to Le Gentil) on behalf of the French Academy of Science, Le Gentil, already a renowned astronomer and discoverer of several nebulae and star clusters, set off in March 1760 to the French colony of Pondicherry in India. Upon reaching Mauritius in July 1760, he heard of the outbreak of hostilities between Britain and France that became known as the Seven Years' War, stranding him in Mauritius. By March 1761 he obtained passage on a ship bound for India but the ship was blown off course by the monsoon winds and had to return to Mauritius. The transit occurred on 6th June with the ship still at sea, so no good observations could be made due to the roll of the ship and the uncertainty of its global position.

Nonplussed, Le Gentil decided to wait it out in the tropics until the 1769 transit. He spent his time mapping the east coast of Madagascar and observing local customs and rites before heading to Manilla in the Philippines. However, he encountered a lot of hostility from the Spanish authorities there and decided to return to Pondicherry, which had now been given back into French hands by a peace treaty. Arriving in March 1768 gave him plenty of time to study the area and erect an observatory but on 4 June 1769 the sky was overcast and Le Gentil saw nothing.

Leaving Pondicherry with dysentery and feeling decidedly under the weather, Le Gentil's ship was caught in a storm and he was dropped off at Reunion Island in the Indian Ocean whilst the ship underwent repairs. He managed to get a berth on a Spanish freighter and eventually returned home to Paris over eleven years after he set off, to find that his extensive letters home had all disappeared due to shipwrecks and enemy action during the Seven Years' War. He had been declared dead, his wife had remarried and his relatives had looted his estate![3]

His partner in observations at the Academy of Sciences, Jean Baptiste Chappe d'Auteroche went to Siberia where, on the day of the transit, he was attacked by the locals who thought his instruments were interfering with the Sun. He had to be protected by a screen of friendly Cossacks whilst he recorded the transit under excellent conditions. He repeated the feat in 1769, this time going to the Baja peninsula (in modern day Mexico) and having a relatively uneventful time until the whole expedition was infected with yellow fever. Chappe d'Auteroche stayed behind to tend to the worst affected and in turn became infected himself and died in August 1769.

Such endeavours demanded the highest commitment of men, resources and materials, with no absolute guarantee of a favourable outcome.

Joseph and the transit of Venus

These events were of course unknown to Joseph, but he was determined not to be left out of this first international attempt. Leaving the grime of London behind, he hoped that he could observe this event from the environs of Trefecca and establish the hamlet's longitude in relation to Greenwich. He arrived on 28th April after dispatching three boxes of mathematical instruments, one of which contained several clocks whilst the other boxes, heavily wrapped in matting, contained a five-foot focal length telescope and possibly a quadrant of some kind.[4]

He was the only observer to watch the event from Wales.

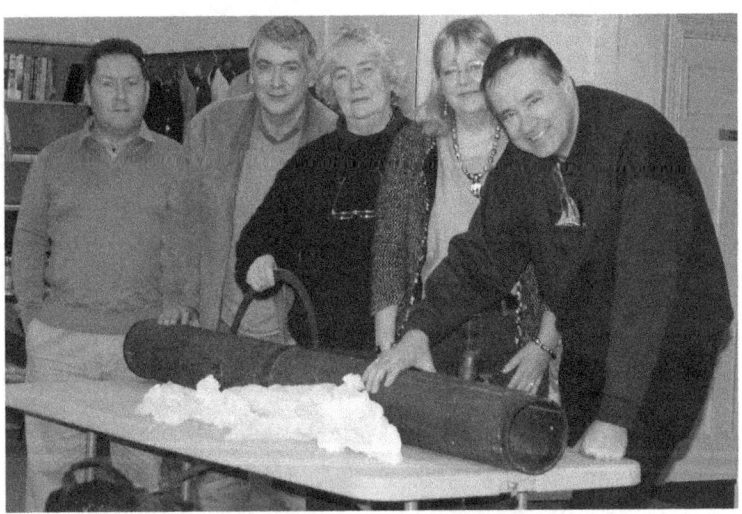

FIGURE 18 Joseph's telescope under investigation
(photograph by author)

Joseph's five-foot focal length telescope laid on a trestle table with protective tissue underneath the tube. The author is on the far left, Jenny Stanesby-Moody is at centre, with Professor Peter Duffet-Brown on the right. Flanking Jenny are Richard and Mair, the secretaries of Trefecca College.

Coincidentally, the transit was not the only event Joseph observed whilst at Trefecca. On the evening of 17 May 1761 a total lunar eclipse was visible and Howell and Joseph stayed up until 1 a.m. watching it. The incidence of greatest eclipse would have happened at 22:11 Universal Time (UT). The moon would have been in the constellation of Scorpio, relatively low on the horizon and sunset would have occurred over an hour previously. The penumbral part of the eclipse would have already occurred before moonrise but totality would have been visible for 94 minutes. This eclipse was central in that the lunar disk passed through the centre of the Earth's shadow, turning the moon a deep coppery-red.

Whether Joseph and Howell used the five-foot telescope that Joseph had shipped to Trefecca he did not state, but it is more than likely that it would have been used to collimate and calibrate his instruments and clocks by observing the shadow crossing the lunar surface. We will never really know as the only testament to this observation is a diary entry by Howell on the following day. It is a pity that Joseph did not record this occurrence; perhaps he just wanted to enjoy the spectacle with his brother who was shortly departing for Bideford in Devon and subsequently missed the transit.

To establish the longitude of Trefecca properly, Joseph would have to make a meridian line based on local noon. He retired to a darkened, south-facing room over the communal oven at the main building in the community and bored a small hole through a plate fixed to the roof. Then, he inscribed a point on a wooden plank and measured its accuracy under the hole with a plumb line, then drew concentric circles around the central point and, as the sunlight penetrated onto the plank at local noon, drew lines marking the solar passage across the circles and noted how the light bisected the circles through the centre point. This method then determined the local meridian and enabled Joseph to set an accurate time on one of his clocks reflecting local noon. The difference between noon measured at Greenwich by one watch and local noon at Trefecca measured by one timed with the meridian line would then give him an estimate of longitude. It appears that Joseph also used the transit timings and Halley's parallax tables to determine this initially and then check timings on his return to London.[5]

FIGURE 19 Total lunar eclipse (photograph by author)
The moon in total eclipse. The moon is full but is stained red by the passage of sunlight through the atmosphere.

His record of the transit shows that he had some difficulty in being prepared for this event. Although he had a good telescope, he records that he had a problem timekeeping as his clock gained 53 seconds in a 24-hour period. Also, the weather was unfavourable, delaying his attempts in setting a proper meridian line to make accurate measurements until two days prior to the event. One is presented with the picture of an old man hastening to make the best he could of an increasingly bad situation between the weather, his instrumentation, perfection in measuring technique and his own health. He reports that on 6th June he 'saw the sun rise and it continued clear the whole morning until past noon. But not being then able to bear much fatigue, I confined my observations chiefly to the times of the two contacts with the sun's limb'.

These were the third and fourth contacts when Venus had already transited the Sun and was now in the last stages of the event. Joseph

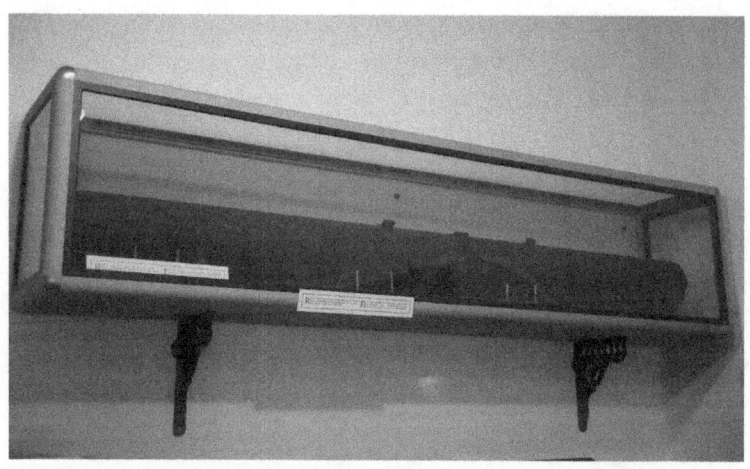

FIGURE 20 Joseph's telescope in a glass case at Trefecca
(photograph by author)

Joseph's five-foot telescope, one of the instruments used to observe the 1761 transit of Venus. The telescope is in a glass case and is mounted high on a wall of the Trefecca Museum.

timed the difference between third and fourth (final) contact to be 18 minutes and 15 seconds. His notes imply that he used the telescope throughout to observe the event, though he does not mention whether the image was projected or seen directly in some fashion. Joseph also admits that he did not take the Sun's altitude either. Although he confesses to having a good instrument for that purpose, he omits what sort of instrument this was, though it was probably the quadrant. Mindful of latitude and longitude, Joseph reported hopefully to Howell in a letter in November 1761 that his observations and calculations would settle a point on the map as to the exact location of Trefecca so that it 'may serve as a standard by which to reckon the exact geographical situations of neighbouring places'.[6]

Upon returning to London, Joseph ascertained the Greenwich timings of the transit event to check the accuracy of his original London timings and confirmed his work on the longitude of Trefecca. He found that there was a difference of 13 minutes and 42 seconds in the timings between each place and from this he calculated that the longitude of Trefecca was 3 degrees 25 minutes and 30 seconds west of Greenwich. He communicated these results by letter in November to George Parker, second earl of Macclesfield, then President of the Royal Society.

It is possible that these observations were used as part of James Short's computations of both the transit and possibly the Solar parallax. The Sun's mean distance from the Earth was the key problem in determining distance scales in astronomy and the Solar parallax was a secondary choice to establish this distance after the transit observations. To perform these calculations one would have to know the equatorial diameter of the Earth and the equatorial diameter of the Sun and its apparent centre on the photosphere. Given that Joseph was not near the Earth's equator, his calculations may have not made much impact on this problem. Astronomers were still refining the Earth–Sun distance based on Solar parallax right into the twentieth century, so it is unlikely that Joseph observed the transit with such problems in mind.

The letter Joseph sent to the Royal Society is still at Trefecca college in facsimile and I render it here as it appears. However, I have to give due credit and thanks in that the work of deciphering the letter was originally done by Jenny Stanesby-Moody of Tredustan House and the letter appears along with her explanation in the magazine *Brycheiniog*.[7]. Although I have also perused the letter and made my own copy, indeed was consulted by Jenny on some areas outside her expertise, Jenny's account is worth placing here in recognition of her fine work and with her permission. The original letter can be found in the National Library of Wales archives as it was left with Stainsby Alchorne to deliver to Macclesfield. Only the facsimile of it is at Trefecca college. The original letter was donated to the National Library of Wales in December 1960 by the Talgarth Historical Society. It is also worthy of note that Jenny is a direct descendant of Stanesby Alchorne, Joseph's assay assistant.

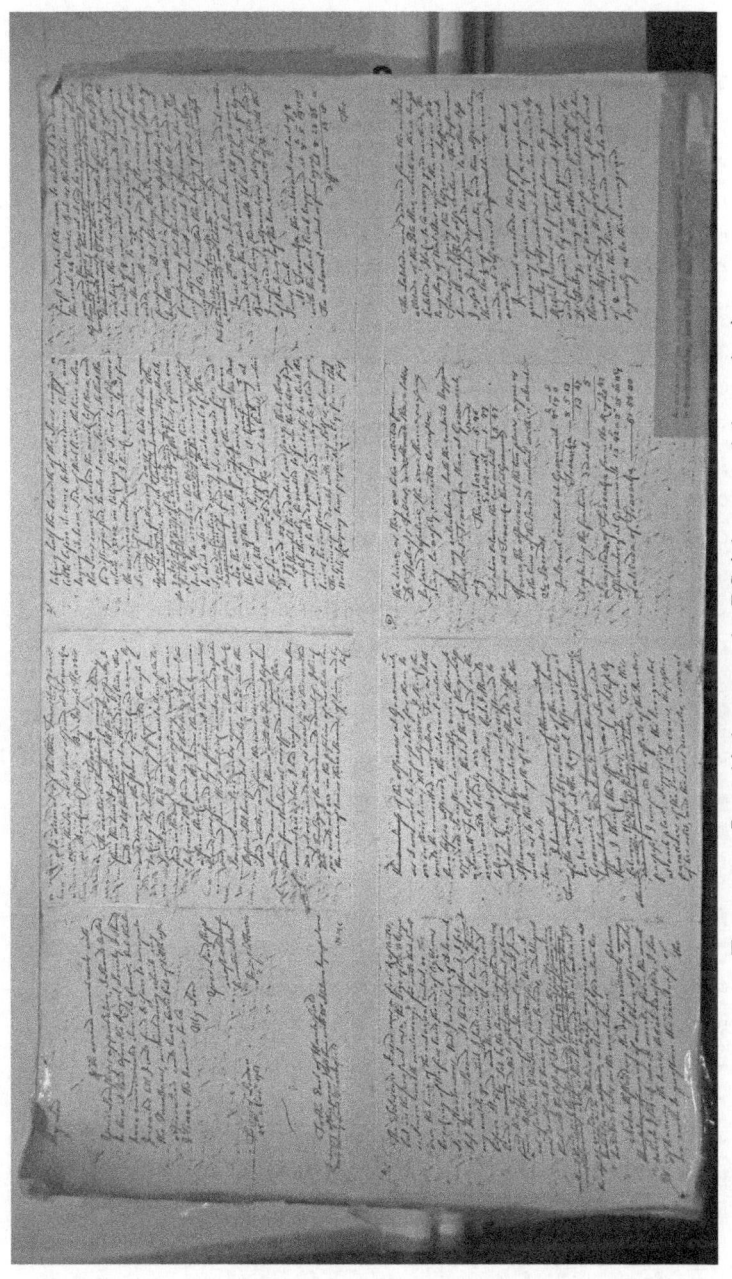

FIGURE 21 Joseph's letter to the RS (photograph by author)

Large board with the pages of Joseph's letter to the Royal Society with his observations of the eclipse laid out at Trefecca college.

Joseph's record as written

The weather at Trevecka proving cloudy from the time I got there till the 5th of June, I could not till then get a Meridian line: This was drawn on the floor of a darkened room, by taking the Suns image projected through a small round hole through a metallized plate fixed in the roof. All the necessary cautions were taken in the finding of this line: A point, on a firm block, exactly under the hole above mentioned, was found by a plummet having a conical end; and from this point as a centre was described several concentric circles upon a smooth plank, before set horizontal and also level with the said centre; and from the near coincidence of a line drawn from thence, with the several bisections made from the several correspondent points in the concentric circles, I had reason to conclude that we got our meridian as exactly as this method and the size of the room would admit of.

I think that we could not err in the position of our line above the value of two or three seconds of time, and by taking half the breadth of the Suns image a little before it came to the meridian line, and laying it to one side of that line, the time when the Suns image touched the mark left there, could be distinguished to about one second; so that the whole error in taking the Suns transit over the meridian could not, I think, exceed about four seconds of time.

The two following days we had the Sun clear at noon, which gave us the rates of going of the clock and of my Stopwatch, the equalling of that time being allowed for. In the following account of the times of the contacts, the error in the meridian line amounting to about 4 seconds, arising from the increase of the Suns declination during the interval of the time taken in finding it, is allowed for; and also the error in the going of the watch from the time of the internal contact of Venus with the Suns limb, till noon, supposing it went at the same rate as it did the next 24 hours, in which time it gained 53 seconds.

I thought this detail necessary, that others might thereby be enabled to form the better judgement as to the degree of exactness, to which the times hereafter mentioned may be relied upon. The principal doubt with myself is, about my Watch keeping time proportionally from the first contact till noon, to what it did during the next 24 hours.

But as the Watch was a good one, and the interval it had to err in was but less than four hours there is no great room to fear that any great error could arise from thence. I have regretted since that I did not take the Suns altitude immediately after the transit of ♀* was over, which would have given me the time to a great exactness, as I was provided with a very good Instrument for that purpose. But being then in a weak state of health, without a proper assistant, and not foreseeing that that use might be made of my observations which I afterwards thought might be, I omitted the taking of that little additional trouble, an omission which nothing but the causes above assigned could have excused.

June 6th 1761. I saw the Sun rise, and it continued clear the whole morning till past noon. But not being then able to bear much fatigue, I confined my observations chiefly to the times of the two contacts of Venus with the Sun's limb.

At Trevecka the internal contact of Venus with the Sun's limb happened at

	h m s
	8: 5:13 a.m.
The external contact or final egress at	8:23:28
Difference	18:15

The Telescope I used was a five foot Reflector; but in the present case, the size of the telescope is scarce worth mentioning; for with that small one, the time of the internal contact or the breaking of the fine lucid thread of light was so instantaneous, that I was sure of it to much less than a second. At that instant I set my watch which I held in my hand a'going, it being before stopped with the minute and second hands set to the beginning of the divisions; and thus, I found the length of the interval of time from noon.

---

* Symbol for the planet Venus.

But the time of the external contact I could not ascertain to the same exactness; herein I was doubtful to three or four seconds; and I suspect that with the best of Telescopes, this observation could not be ascertained to any great precision; and without the utmost attention a very considerable error might be easily committed: Nor is this to be wondered at, when it is considered that in the present case any given error in the observation as to space, will be magnified above two hundred times in the conclusions.

Notwithstanding the disagreements between many of our home observations, which I think arose merely from the want of having the times better adjusted, I see no room to question the exactness of the observers at Greenwich, as it was next to impossible for them to err in their time; the only error, I think, that could be committed anywhere. For, as hath been before observed, the internal contact appeared so instantaneously, even through a small telescope, that I think the greatest novice could hardly err one second in the taking of that observation; but I should not wonder if persons not accustomed to astronomical observations, should differ much as to the length of time between the two contacts.

From the observations of the internal contact, made at the Royal Observatory at Greenwich, and at Trevecka I think the difference of longitude between these two places may be safely obtained within the limits above mentioned. For this purpose, I computed the effects of the parallaxes at each place; supposing the horizontal parallaxes of the Sun and Venus, the position of her orbit, and the Suns diameter, were at the time, as they are to be collected from Dr. Halley's Tables; and should these tables be found defective, the error thence arising may be easily corrected hereafter.

By my calculations both the contacts happened sooner at Trevecka than at Greenwich, viz.

	seconds
The internal	5.46
External	1.99
Duration between the two contacts longer at Trevecka than at Greenwich	3.47

Hence, the observers at the two places, agree as to the time of the second contact within about 2½ seconds.

	h	m	s
Internal contact at Greenwich	8	19	0
Trevecka	8	5	13
Difference		13	47
Neglecting the fraction, deduct			5
		13	42

Longitude of Trevecka from the Royal Observatory at Greenwich

                m   s
                13 42 = 3°   25m 30s W'ly long.

The latitude of Trevecka   51°N 58m 30s

The latitude was deduced from the meridianal altitude of the Pole Star, which in these high latitudes I take to be a very good method, the long stay of that Star upon the wire in the Telescope, giving the observer a sufficient time to verify his observation. The Instrument I used I could depend upon to rather less than the ⅓ of a minute; and two observations made at different times coincide exactly.

I cannot conclude this paper without giving my opinion, that if a competent number of observations have been made by skilful persons at proper places, the great end proposed by our late great Astronomer Dr. Halley, may be obtained perhaps to the degree of exactness mentioned by him, notwithstanding the position of the track of Venus over the Sun, proved not so advantageously as he then imagined.[8]

It is unknown how much of Joseph's observations were used by James Short as usually a whole transit has to be observed with accurate timings to be of any use to those making the computations. Joseph could not have seen the entire transit anyway as the first and second contacts with the solar disc occurred whilst the Sun was still below the horizon before sunrise at Trefecca. It would seem that Joseph wanted to accomplish as much as possible and report on a singular event in the hope that his observations could be of some use.

Having observed a Venusian transit myself in 2004, it is impossible not to realise that it is a once in a lifetime event, not to be missed at all if one is interested in astronomy. No doubt Joseph felt the same and took advantage of the opportunity and also perhaps he realised that he may not get another chance at such a singular event given the decline of his health.

Together, the observations of the lunar eclipse and the Venus transit probably marked the last time that Joseph was active in the astronomical field. The fact that the telescope still resides at Trefecca and in all probability some of his clocks were also left there suggest that he had no further use for them, or, at least, recognising his own failing health he left them behind to be of some use to the Trefecca community. Nevertheless, his establishment of the Trefecca meridian and geographical location was important and his timepieces were used for some decades hence as a standard of local time.

JOSEPH'S ECONOMIC INFLUENCE AND ESSAY ON MONEY AND COINS

Money as a form of exchange has a long history. Across Europe money became useful for the simple reason that although merchants could deal with each other by issuing notes of exchange or IOUs and then clear their debts by paying off the IOUs by using recognised monies at meetings in country fairs or at regulated times, the system of IOUs would only work if everyone was completely trustworthy. Money therefore became a social institution that solved the problem of trust. Even money in today's economies is a form of IOU with well-known and regulated values. Just look at the back of any denomination banknote: 'I promise to pay the bearer on demand the sum of xxx', albeit in very small print. Money is therefore the trustworthy means of exchange and people are happy to accept it in exchange for goods or services.

Although money was useful as a token of exchange, we have seen that what was actually being exchanged could vary. Far from there being a uniform and coherent system, local anomalies and customs had over time created many disparities, complicating internal and external trade. Beside money, weights and measures as forms of transaction also suffered. It would seem ludicrous to us today that a weight such as a pound (lb) could be the equivalent of 16 ounces (oz) in one area, yet a few miles away it could be 32 ounces. Even the weight commonly used in the imperial system today in the UK such as the stone (14lb) could in some places across the UK be as much as 16lb or 20lb. Naturally, it was time to look at the whole system, review it and revise it accordingly as no real revisions had been undertaken, despite many attempts, since the time of Queen Elizabeth I and before that King Edward I.[1]

Adjusting these standards was not Joseph's initial role. However, his duties as Assay Master ensured that his opinions and expertise made him available to royal circles as well as to the government of the day. The introduction of monometallic ideals and the gold standard under Isaac Newton made the Royal Mint the first institution that anyone could turn to and accept as the revisionist of standards, not just of coinage but also of weights and measures, which had by the middle of the eighteenth century fallen into a parlous state. The responsibility for weights and measures was held to be a crown matter and would come under the jurisdiction of the Royal Mint. As we have seen, Joseph's contributions to the Carysfort commission ensured that standards of weights and measures were addressed.

As Assay Master, Joseph would have had to know about the purity of metals, the weight of each metal and the alloys that could be used to make coins. His duties also extended to making weights for merchants so that a standard could be achieved, at least locally in London, and that standard could be extended to the entire country. What attempts were made to solve the problem of discrepant weights and measures, and how successful were they? The problem must have been significant, because between 1757 and 1765, during which time Joseph and his assistant Stanesby Alchorne were employed at the Mint, the government oversaw two very important select committees and four failed bills. Joseph's attempts to standardise weights and measures set the Mint on the road to a fairer system and his influence, even after his death, had massive implications: the government statute of 1824 introduced the imperial system that until recently was in use in the UK and in some measure is still operational in the USA.

Similar standards now had to be applied to coinage. On the face of it this seems a simple enough task for an Assay Master, but nothing could be further from the truth. Part of the problem was money coinage itself. In Britain since the sixth century, coins themselves were almost nothing more than tokens of exchange. Some scholars, such as Nick Mayhew, suggest that medieval coins were nothing more than gift tokens rather than a medium of exchange. He examines evidence which is suggestive that monies in the form of coins were overshadowed by other methods of exchange such as barter, or the provision of labour for a given task or time. Though coins of silver and gold were in circulation, 'pounds', 'shillings' and 'pence' were neither common parlance nor used to designate goods to a specific value.[2]

However, other evidence seems to suggest that coins were available across medieval Britain in their millions and were minted in various locations as and when the need arose. There was a simple equation between the amount and value of money in circulation and the prices of goods but fluctuations were difficult to predict and prices harder to control without a set value system. Economically, this ideology has been named Quantity Theory and it still has a hold over economic policies today. The amount of money in circulation has an effect upon prices, upon inflation and upon interest rates. Flooding the market with money has been thought to be disastrous, though modern ideas such as quantitative easing during the banking crisis of 2008 introduced more liquidity to the markets and prevented a rush on the banking system.[3]

In Joseph's day, the amount of bullion available to countries would often have a major impact upon its money and coins. If the bullion supply was low then more money would be minted from a limited amount of bullion, making the coins thin, friable and with a perceived smaller market value. Conversely, more bullion resulted in fewer coins but with better regulated rates of exchange and consequently higher values as they had larger gold contents and were consequently thicker. To determine the exchange rate between countries and to fix the value system in your own country to coinage was a risky business that would affect the economy adversely. Fixing the value of coinage and maintaining it was very difficult and had to take into account fluctuations against foreign currencies and exchange rates that would influence import and export markets.[4]

Royal Mint records in the UK show that the number of coins in circulation between the thirteenth and eighteenth centuries rose and fell as some were recalled and recoined, soldiers in wars on the continent lost them or took them abroad as exchange tokens, or existing systems became obsolete and new ones were introduced. In addition to this, currency underwent loss or wastage at a rate of about 4 per cent a year and the value of the metals changed markedly over the centuries as silver and gold waxed and waned in use and value. In addition, the use of copper in coins of low value, but important in the economy for the use of poorer elements of society, meant that the coins were of no use against the values of gold or silver but were seen as useful tokens of exchange or 'commodity money' in the words of Adam Smith, with values that changed over the centuries.[5]

What was really needed was a standard for all coinage and its tethering to a fair and rational system of exchange that would stand the test of time and also be accepted as a means of exchange in foreign markets. Money had to become a 'store of value' and a proper unit of account and a reliable and stable medium of exchange unless influenced, not by circumstances, but by government policy. Even before Joseph's day, the economic system was moving away from dependency on subsistence and agriculture economy to an exchange-based system. John Conduitt, who had been Master of the Mint and recommended Joseph for the assay master post, had already written extensively on maintaining a standard that was tied to the value of gold and on the values of gold and silver in England in comparison to other countries. He argued that the values should move closer in line with other trading countries so that fairer rates of exchange could be obtained.[6]

An Essay upon Money and Coins

Joseph took this problem head on, but it was apparent to him that the value of coins could only be guaranteed if there were standards that were universally applied. In order to understand and ascribe such a standard, he wrote *An Essay upon Money and Coins* (part I), referred to as Essay dedicated to a former Master of the Mint, Richard Arundell MP, who became Lord Commissioner of the Treasury and Clerk of the Pipe Rolls wherein government expenditure and income were recorded. Arundell was also the brother-in-law of Henry Pelham, the prime minister from 1743 to his death in 1754 and his dedication to the work reveals the company of great men that Joseph now enjoyed. Arundell died in 1758 and was lamented by Joseph in the preface to *Essay* when he said that Arundell 'intended amongst his other great designs for the good of this country to have made such regulations in regard to our coins as would obviated all complaints about them in future'. It became Joseph's intention that *Essay* would not just prescribe such regulations but underpin them with solid economic ideas.[7]

Essay demonstrated that coinage had to be tied to economic stability and the sources of wealth in such a way that money became the standard measure due to its inherent trustworthiness. He also recognised that labour and rates of labour payments were the foundation of any economic system and a just wage and encouragement to work should be part of any country's economic goal. Looking ahead, he maintained that a welfare state should be set up in order to aid families whose breadwinner was temporarily unemployed or unable to work due to disability. He also recommended that banks should be regulated so that there were no shocks to the market, the former recession after the South Sea Bubble debacle no doubt in mind.[8]

Joseph saw that there were some jobs that required the application of education, certain skill sets and intelligence, which should demand a higher wage that set them apart as professionals in their field, a circumstance that he hardly needed to defend as this had been the case for centuries. Now it was being placed within an economic system that recognised the hierarchy of labour as part of professional rewards.

Joseph also realised that minted coins in the past had decreased in metallic value even though its holders regarded the face value of the coin to be as it was displayed. This formed the basis of trust in coinage despite weight reductions in the past and the amount of bullion included in each coin. In the past, the medieval government of England had reduced the weight of silver in coins from 22 grains of silver to just 11 grains, cutting by half the amount of precious metals yet retaining the trust in the face value of the coin. There was no law against doing so as long as such agreement was recognised by monarch and parliament, but this possibly did not do so well against foreign currencies despite the fact that home prices remained unwavering in the face of the reduction. Joseph recognised that such debasement of the metallic value did not especially affect the value of the coin as long as there was universal assent to the face value.[9]

To achieve such trust, coins would have to be minted from metals that had regulated rates of exchange, that is, gold and silver. Occasionally, to pay off existing debts, monies with lower metallic inclusion would be used, especially to pay foreign debts, but the imbalance of value in this practice was recognised and stopped by most continental courts. Therefore, money across a wide range of circumstances and services had to be trustworthy to all and so Joseph advised in *Essay* part II that across Europe both gold and silver in standard weights be recognised as true universal values, with silver being the best arbiter of this value as its market value wavered less than the market values of gold. Hence, across all foreign and home exchange, silver especially but gold as the standard should be the centrepiece of coinage.

It became Joseph's duty to try and achieve this. To produce a fair standard, he eventually patented an instrument for the testing of coins, probably a form of money balance that ensured that each minted coin was monometallic, ensured accuracy of individual weight and adhered to strict regulation. Typically of him, Joseph refused to be named in the patent, which was retained by the Royal Mint. In this way a guarantee of trust was established in accurate coinage. He further reinforced this issue by writing part II of *Essay* subtitled: 'wherein is shewed that the established standard of money should not be violated or altered under any pretence whatsoever'. This second book was dedicated to the Chancellor of the Exchequer, Henry Bilson-Legge, indicating again the high political offices now taking Joseph's advice. Within each chapter Joseph reveals knowledge of general philosophy and a penetrating insight into its political and social application in that he studied John Locke's ideas on Monometallism and applied them to the gold and silver standard.

His last commentary was upon the banking system and the need for regulation that would control speculation and thus weather any economic upsets. He also called for a just standard being applied to bills of exchange; if there was any fluctuation in the value of silver or gold then foreign debts would need to be paid at the value rate of the debt of the time, not base it on the value of silver or gold at the time of payment. This takes into account that fluctuating levels of silver and gold values would even out and a fair trading policy could operate without either party becoming indebted or fearing losing out on value or payment.

That Joseph used sound reasoning and had a very good grasp of economics and its impact upon the country is evident when one considers that many of the conclusions reached in part I of *Essay* presaged the information used later by Adam Smith in his epochal 1776 volume, *An Inquiry into the Nature and Causes of the Wealth of Nations*. Familiarity with Joseph's economic concepts is quite clear within the text: Smith referenced the *Essay* in Book I, Chapter I of *The Wealth of Nations*, which borrowed substantially from similar arguments to those Joseph had written nineteen years previously. Such subjects as the division of labour, wages for labour, the origin of money and productive and unproductive labour were no doubt subjects discussed by Smith and Joseph, who knew each other from their fellowship at Boodles gentleman's club and no doubt exchanged ideas frequently during their conversations.[10]

Shortly after Joseph's death in 1764, part III of *Essay* was printed. The writer was Stanesby Alchorne, the assistant Joseph had trained who now became assay master in his turn. Alchorne combined the notes and added pieces to it from what he recalled from many conversations on the subject with Joseph and from his own observations after perusing the manuscript Joseph left behind. It deals with remedying the bad state of the current coinage and, again, redressing the trust the economy should have in coinage standards.[11]

Although much of what Joseph observed in *Essay* was already in large measure part of the economic systems around him, his observance of standards and fairness shine through the work; and tying coinage, bullion rates and exchange to universal standards and metals with regulated value revealed the economic grasp and intellect of this man, a former son of a blacksmith and smallholder. That *Essay* was dedicated to the highest treasury authorities of the land is testament to his standing and regard by such luminaries.

Today, *Essay* can still be found on the bookshelves at the Houses of Parliament and is now available in print through the *Cambridge Library Collection*, a resource boasting that its collection is of 'enduring scholarly value' and the titles are 'important for professionals or researchers for the source materials they contain or as landmarks in the history of their discipline'. Joseph may have been surprised yet pleased that his contribution to the history and economy of our nation would receive such attention.

10

JOSEPH'S FINAL SCIENTIFIC TREATISE

That Joseph maintained a keen interest in the science of his day and was on familiar terms with many in positions of political power and the Royal Society is evident from his letters, his Parliamentary consultancy and the incidental writings of his contemporaries. Over the period of his employment at the Mint, Joseph had amassed information, ideas and materials on the subject of optics since at least 1742 and was working on its final layout just before his death.

Although Joseph had initially divided the book into compact sections, the notes and information he had collected were a little disparate and were placed in the best order available to be published. As its compiler admitted in the introduction to *Treatise of Optics*, the materials had been considered by 'some gentlemen well acquainted with the science of optics' but the greater part of the work could not be properly arranged without the assistance of its author. Despite the attentions of those 'gentlemen', undoubtedly members of the Royal Society, the book always remained incomplete.[1]

A Treatise of Optics

Published posthumously by the publishers, B. White of London in 1775 as *A Treatise of Optics*, it seems evident from the text that Joseph initially wanted to limit his optical work to microscopes, but as he grew more and more interested, the more it expanded into examinations of biology, telescopes, camera obscura, lenses and the metals used for mirrors.

A Treatise on Optics is rendered in two parts: Book I deals with elementary optics, lenses and the properties of light, whilst Book II returns to his experiments on vision and greatly enlarges upon his earlier experiments. His technical expertise enabled Joseph to describe the construction and use of a camera obscura to aid drawing and painting and he included the construction details of a 'pocket camera' with a 1.5-inch lens that could be fitted to the handle of a cane in order to make a portable instrument for landscape painters. It would appear that his ability to design devices suited for specific or scientific purposes was undiminished despite his duties at the Mint.[2]

His interest in astronomical observation led him to write about stellar magnitudes (the brightness of stars), the transparency of the atmosphere and the aberration of starlight due to the movement of the Earth around the Sun. This latter was especially pertinent after the discovery of this phenomenon by the Astronomer Royal James Bradley, a discovery that finally established the heliocentric theory of the Sun-centred Solar System. Joseph had already included this Sun-centred factor in his earlier work *The Description and Use of the Globes, and the Orrery*.

Joseph starts Book I with an introduction detailing the properties of light. The speed of light had been originally calculated by the Danish astronomer Ole Roemer in the seventeenth century but after further calculation the speed was revised to its more modern value of 300,000 kilometres per second (186,240 mph) by astronomer James Bradley. After this brief introduction Joseph delves into the properties of light as they reflect and refract depending on the medium they encounter. He then goes on to describe the properties of lenses and how they form images. Throughout these descriptions, Joseph posits mathematical exercises that enable the reader to understand such things as the refractive index: determining how light bends when encountering such media as water, glass, space, air and other gases. Some of the ideas he discusses may be familiar from school physics courses and are still part of many curricula today.

He makes the observation about our Earthly atmosphere that 'though it is the most transparent medium we know, it is full of opaque particles that impede light', recognising that the opacity of the atmosphere will affect anything we observe through it. Joseph draws attention to the faintness of stars near the horizon as their light is absorbed by our atmosphere and proposes that allowance be made in the calculation of stellar brightness due to this absorption. Two centuries later, astronomers discovered that stars in our galaxy also have to have their brightnesses corrected due to the materials in the Interstellar Medium, the gas and dust between the stars, that also absorb their light. Much of Book I is technical and quite demanding to read if one is unfamiliar with optical concepts. Nevertheless, his methodology lays a good groundwork to understand the images that result from using different optics and lenses and how they would appear to the user.[3]

Discussions on vision

In Book II of *A Treatise of Optics*, Joseph made a foray into the limits of the human eye by delving into biology, using experiments to ascertain the optical structures of the eye and measuring its resolving power. He wanted to determine what the minimum visibility of the human eye could be and how vision can be affected by long and short sightedness and how spectacles would make a difference to those suffering from these problems. Again, he cites mathematical problems that the reader can follow to see how his conclusions are arrived at and presents the latest theories of his day of why the eye should suffer such problems. These theories seem quaint and perhaps antiquated in comparison to our modern knowledge, but we have to remember that this was at the cutting edge of knowledge during Joseph's era.[4]

Joseph determined by experiment and measurement that the diameter of the eye was $9/10$ of an inch (23 mm) and the distance from the posterior surface of the lens and the retina averaged $6/10$ of an inch (16.36 mm). From this he worked out that the tiny angle of 1 second of arc ($1/3600$ of a degree) subtended to $1/10,000$ of an inch (0.0048 mm) on the retina, giving one of the most accurate estimates of visual resolution then available. From his astronomical training, Joseph would have recognised the arc second as a tiny angle and quite impossible to see visually on a heavenly body such as the Moon or a planet. Today, the best estimation of human visual acuity is one arc minute or 60 arc seconds, sufficient to resolve large craters on the Moon under ideal seeing conditions. Only the increased resolving power of a telescope would enable small angles such as arc seconds to be identified.[5]

Joseph mentions experiments he performed on his own eyesight and their acuity by setting up two candles at a 'determinate distance' and then covering each eye in turn with paper and, eventually, a book. He also covered one candle at a time in this manner to ensure that the experiment was as fair as possible. He notes that one of his eyes was a little more light sensitive than the other by just $\frac{1}{13}$ part, though how he measured this is open to interpretation. He also worked out that if he stood twice the distance away from his light sources then they appeared dimmer by a quarter, thus showing that light followed an inverse square law. If one was twice as far away from a source, the light would be one quarter of the original output; four times further away and the light would only be one sixteenth as bright.[6]

He makes the association between brightness of objects and how movement of the eyes can make an object such as a star or the planet Jupiter seem a little brighter. He gave no account of how this phenomenon occurs. Today we know that the cone cells in the eye, which surround the rod cells directly behind the pupil, are a little more light-sensitive than the rods. In visual astronomy we recommend using the 'corner of one's eye' or averted vision as it is known to see fainter or finer detail in distant objects through the telescope.[7]

Once again, leaving future readers to ponder over how he arrived at these conclusions, Joseph does not mention what further experiments he undertook on the eye, or whether they were carried out by other people. Considering the circles he moved in, it is possible that his foray into biology was assisted in some part by James Jurin, but exactly how much collaboration there was between Joseph and Jurin is unclear. Joseph would have been aware of Jurin as the latter was the secretary of the Royal Society up to 1727 and then became the president of the Royal College of Physicians right up to his death in 1750. Even if the two never conversed, Joseph would have been aware of Jurin's extensive experimental accounts on the anatomy of the eye in the *Philosophical*

Transactions of the Society. This possibly points to Joseph's regular attendance at meetings, though he never reveals as much. Additionally, he could have read Robert Smith's 1738 book *Opticks*, which also included an addendum by Jurin.[8]

Joseph also conducted a series of experiments with a chequerboard of squares one tenth of an inch across and also used a number of horizontal and vertical gratings made of fine wires to ascertain the limits of resolvability and the visual acuity of the eye. Such experiments had been done independently by Tobias Mayer at Göttingen and by Jurin in London.

He also gave instructions on a 'perspective box' that could hold a miniature landscape in such a way that the observer would see people through it as if positioned against a distinct romantic backdrop, almost an artist's studio in miniature. It is not known if he actually made any such instruments personally; in all probability his duties at the Mint would have overshadowed many of these optical interests. He records in Book II of *A Treatise of Optics* a description of such a box constructed by a Mr Parrat with the dimension of 16 inches in length by 18 inches wide by 24 inches deep. It is unclear from the text whether this box is merely being described or if Joseph actually had one constructed for him. After all, he had access to all the materials necessary and had the expertise, or knew enough experts, to produce such an instrument.[9]

Telescopes, microscopes and camera obscura

Book II of *A Treatise of Optics* contains a wealth of materials on the camera obscura and pinhole cameras (merely visual aids, no film) and the images obtained with them. Again, Joseph leaves one in doubt as to whether he constructed such instruments or was simply familiar with their use. Joseph's discussion on telescope optics dominates much of the work and despite his intention of considering microscopes only in the *Treatise*, they are only mentioned a few times in the text and that in relation to the images obtained and their brightnesses from using speculum mirrors (see below).

Telescopes in the eighteenth century were crude by today's standards. However, there were some excellent telescope makers in London during Joseph's time there, who constructed both refractors using glass lenses and reflectors after the manner of Isaac Newton's reflector telescope demonstrated to the Royal Society in 1668. Later in the century, telescope makers such as James Short, Jesse Ramsden, John Bird and Charles made some of the best telescopes of their day. Jesse Ramsden would go on to win the Royal Society's Copley Medal due to his excellence in construction of a variety of instruments.

Ramsden would no doubt have come to Joseph's attention when he was apprenticed as a mathematical instrument maker in Fleet Street in 1758 and worked for Sissons and the telescope maker John Dolland before setting up his own business in Piccadilly in 1762. Joseph and Jesse had something in common: just like Joseph, Jesse Ramsden married the daughter of his mentor.

James Short was already known to Joseph (and his letter on the Transit of Venus was used by Short in his computations) as Short had been elected to the Royal Society in 1737 and attended many Society meetings, probably including Joseph's demonstrations of the globe in 1739. His telescope workshop in Surrey Street turned out over 1,300 Gregorian-type telescopes during his tenure there. Short also introduced the use of speculum metal, an alloy of copper and tin, into the construction of telescope mirrors. The reflective properties and determination of focus and focal lengths of speculum mirrors were discussed by Joseph in Book II of *A Treatise of Optics*. Once again, Joseph includes mathematical exercises to calculate such lengths.[10]

Joseph recommends that speculum mirrors have to be exactly figured or ground to a perfect sphere or parabola depending on which instruments they would be used for. A Newtonian telescope has a parabolic mirror which brings the rays of light to a focus. This focus is interrupted by a flat mirror which then sends the light out of a telescope tube at 90 degrees to an eyepiece at the focal point. James Short's Gregorian telescopes contained two speculum mirrors which sent the light up and down the tube three times before exiting through a small hole in the primary mirror, so a long focal length could be established in a relatively short tube. Both mirrors were concave. Later in Book II of *A Treatise of Optics* Joseph discusses the problems of convex and concave mirrors if used individually. He also notes the types of images that would appear in a telescope of each type if one used different eyepieces, a phenomenon that is still problematical for today's visual observers using different telescopes.[11]

Joseph's discussion of the camera obscura centres on the use of lenses and their proper focal length so as to obtain an image that will be of sufficient size. The principles of the camera obscura (Latin for 'dark chamber') had been known for millennia; it is possible that the Chinese had constructed one to observe the Sun in the second Millennium BCE and they certainly had records that go back as far as 500 BCE that demonstrate its use. Even the Greek philosopher Aristotle may have used a camera obscura to observe a solar eclipse. Today the camera obscura is usually a tourist curiosity as one can obtain a panoramic image of the landscape from one by simply stepping into a darkened room with a fitted lens that will project the image onto the ground or onto the walls.[12]

Included in the notes that the compilers arranged within *A Treatise of Optics* are many diagrams that aid the reader in visualising the properties of light, demonstrate how light can be reflected and refracted, and detail what materials can be used to execute this. Joseph's discussion of grinding lenses and mirrors was especially pertinent to instrument makers of his day. Seen from our modern perspective, *Treatise* is an incredible work that reveals a thorough understanding of optics and optical processes and a willingness on the part of the author to stride across different disciplines in order to illustrate, educate and illuminate what could have been a very dry subject. Joseph must have spent many leisure hours teaching himself all about the subject before rendering his experiments and notes in such detail. It is a great pity that he did not live long enough to finish it.

Sadly, the book ends suddenly with no real conclusions, evidently being a collection of orderly notes almost ready for publication, some of it occasionally repetitive. However, its mathematical content is very thorough and drew together much of the known optical sciences of his day, possibly adding much on the nature of the eye, and the way in which vision is perceived via the optics of the eye. It would have been a boon to telescope and instrument makers as its exercises and descriptions laid the groundwork for understanding optics, the uses of reflective metals and the shaping and figuring of glass lenses. Today, *A Treatise of Optics* has been selected as a culturally important book by the United States Library of Congress and takes its place amongst the world's great literature.

11

JOSEPH'S LEGACY AND HIS HEIRS

The impact Joseph's work had on the scientific, political and economic world is generally hard to ascertain as his influence became buried amongst the works of others. He did not help himself as his modesty initially forbade him to put his name to his work and, in many instances, it is only through others that we know of his contributions. That he was respected, his advice taken by those in political and economic office, and that he was esteemed by the scientists of his day is not in dispute. His legacy on weights and measures still influences us today and his thoughts on the properties of money and coins is still significant, his books still in print and his thoughts stretch out to us more than 250 years later.

Still, Joseph seemed to make little of his increasing fame and regard amid his scientific contemporaries. His diffidence even went as far as his instrument work. In an age where men such as George Graham, Thomas Tompion and John Coggs were rightly placing their names on their instruments, as they were as much works of art as useful tools, Joseph refused to do so. Inscribed on his plaque at Talgarth church is the phrase 'superior to the love of fame, forbore even having his own name engraven on them', revealing a retiring modesty and contentment that he was part of a larger body of enquiry into nature. Nevertheless, his fame was recognised in the district of his origin. On one of his visits home, his brother Howell describes how walking down the street with Joseph 'filled me with pride that I was in company with such an accomplished person'.[1]

Today, it is becoming increasingly evident that Joseph's influence on the life of this nation was profound. It is a pity that he wished to stay so much in the background as nothing more than a voice in the choir during the Enlightenment. His observations in the Caribbean and dedication to instruments and timepieces paved the way for the acceptance of longitude based upon chronometers and accurate astronomical observing. His book on navigation and use of celestial and terrestrial globes as aides continued to influence the Royal Navy into the nineteenth century. His star maps were used by the foremost professional and amateur astronomers of his day and paved the way for the discovery of the planet Uranus by William Herschel, who used Harris's northern map to check the position of the planet near the star η Geminorum and later in the production of Herschel's *Catalogue of Double Stars*. Without the accuracy of celestial maps, who could say when Uranus or Neptune may have been discovered? Joseph's cartography, attention to detail and accuracy in positioning is a legacy that should be celebrated.

Joseph once wrote to his brother Howell that 'a willing mind and an upright heart are the best qualifications for any undertaking.' No one knew this truism better than Joseph who had lived a full life in science and public duty with no other credentials. Such was the trust placed in him by the governments of the day that Joseph was selected to be a commissioner in the peace treaty that saw the end of the Quebec wars with the French in Canada and the settlements between the British and native tribes. In the event Joseph declined the honour as he felt that he was 'unworthy of such a position' and was still recovering from a long but undefined illness.[2]

The *Dictionary of Welsh Biography* claims that Joseph was also the author of several anonymous works on astronomy and mathematics, but it appears to be no longer possible to discover what these works are. Without them it is not possible to accurately ascribe any such work to him. Perhaps Joseph regarded such works as minor details, or they may have been exercises that were written up for lectures on natural philosophy to nobles and visiting dignitaries or even for Royal Society meetings. D. J. Bryden notes that after Joseph's return from Jamaica he promised to submit more observations 'made with great care' on magnetic variation, but these never materialised. Perhaps it is these notes that Lloyd and Jenkins refer to or to his books on *Navigation* and *Globes* as initially they did not display Joseph's name.[3]

One form of instrument that Joseph continued to be interested in and had used most of his life was the clock. When his brother Howell established the Trefecca religious family – Teulu Trefecca – in 1754, he supplied several clocks for the community and his personal expertise was useful on visits to Trefecca during their set-up. One such was a main clock with four faces established for local timekeeping, though whether Joseph had any hand in this particular instrument is open to interpretation – local sources claim it to have been set up by Joseph, but this is not definitely established. Nevertheless, the clock became a local curiosity and country squires from many miles around would regularly set their watches by the 'Trefecka Clock'. This instrument, with only one face rather than its original four, is still at Trefecca College in the main hall and still maintains time when wound, over two hundred and fifty years later.[4]

Joseph's interests outside the Mint

Joseph took a keen interest in the life of the Trefecca community but also contributed to the life of his fellow Welshmen in London. He was an early member of the Honourable Society of Cymmrodorion, a cultural society set up by Welshmen in London in 1751, founded by the brothers Richard and Lewis Morris to encourage interest in literature, science and the arts as connected with Wales.

Lewis was a cartographer who had undertaken a hydrographic survey of parts of the Welsh coast, so he and Joseph may have had a lot in common professionally. Additionally, Richard was the chief clerk of foreign accounts for the Royal Navy, with an office close to the Tower of London and no doubt would have encountered Joseph frequently during the course of their daytime jobs and travels to work. Morris writes about a visit Richard made to the newly built British Museum in the company of the 'Tower gentry', which included Joseph. The Lewis brothers were certainly aware of his stature. The brothers were also involved with printing bibles in Welsh for the Society for the Promotion of Christian Knowledge. Richard died in 1779, having retired from the Navy's employ at the Tower of London.[5]

The Cymmrodorion Society still meets in London today and is one of the oldest societies highlighting the traditions of Wales outside the principality. Part of its mission then was to raise funds to assist impoverished Welsh people in London and Joseph no doubt contributed much to this charity. The Cymmrodorion Society, under the leadership of the Morris brothers, was partly responsible for the 'Welsh Renaissance' of the eighteenth century and Joseph found himself in company with some of Wales's great literary figures of the age. The 'Morris letters' refer to Joseph by name as well as that of Howell and indicate his generosity to others and his deeds in connection with the Mint.[6]

Through this society, the bilingual Joseph met the historian and poet Evan Evans, who rediscovered many old Welsh poems by the bards Taliesin and Aneirin from the sixth century. Evans's work on Welsh bards and their connections to historical places and people opened up the hidden treasures of Welsh literature and their relationship to the land of their birth. Joseph also encouraged the poet Goronwy Owen, even contributing a guinea to him when Owen left Britain to take up a teaching post at New Brunswick in the American colonies before becoming a plantation owner.

Joseph's association with the politicians and philosophers of the day continued through his work at Boodle's, a gentlemen's club. Initially the club was known as Almack's as it met next door to Almack's tavern in Pall Mall. It later moved to its current address at St James's Street and was named after its head waiter Edward Boodle. This exclusive club is the second oldest such club in the world and is still in existence today as a place of relaxation, dining and socialising for prime ministers, politicians and aristocrats. Although not politically aligned, the membership throughout its history has favoured the Conservative party. Membership is by invitation and election only and in its initial years was limited to a membership of 250 who had to pay a yearly subscription, which still stands; in its first year (1762) there were eighty-eight paid members. It has become almost equally famous for its 'Boodles Orange fool', an indulgent dessert of oranges, liqueur and cream.

It appears that Joseph was one of its first executives, possibly the secretary for a year as the controlling board rotated the duties of its managers. During his tenure there, its members included the philosopher David Hume, the economist Adam Smith, William Cavendish the fifth Duke of Devonshire (husband of famous socialite Georgiana), the politician Charles Fox, the historian Edward Gibbon and its founder William Petty-Fitzmaurice the Marquess of Lansdowne who would go on to become the prime minister in 1782. Later members included the abolitionist William Wilberforce, Arthur Wellesley the Duke of Wellington (and future prime minister), the socialite Beau Brummell, statesman Winston Churchill, disgraced minister John Profumo and the James Bond novelist Ian Fleming.

Joseph was moving in high society indeed but incredibly always managed to keep his feet on the ground and his head out of the clouds. His diverse interests are very typical of educated gentlemen of the day and these interests were not just confined to London.

On his frequent returns to his old home in Trefecca, as we have seen previously, Joseph gave freely of his ideas in the building and administration of the trades carried on there. Joseph also encouraged the use of new farming machinery in addition to assisting Howell in joining the Brecknockshire Agricultural Society. The 'Trefecca family' were already known for the quality of their wool and the ample foodstuffs they sold at market. The society were also instrumental in setting up Brecon market in 1756 as an outlet for such activities and to cultivate a market for quality products across the county. Thus, Joseph also played a small part in the agricultural processes then assisting the burgeoning Industrial Revolution in which Wales had a central role.[7]

Joseph's heirs and descendants

After the death of his wife Anne in April 1763, Joseph became increasingly frail, complaining of attacks of gout and increasing tiredness. Although he wrote to Howell in April 1764 telling him that he was now in better health and concentrating on writing up his book on Optics, even talking of a recent lunar eclipse and the care his daughter Anna

Maria was giving him. Joseph was already very weakened, and a final illness took him just a few months later. He died on 26 September 1764 and his body was interred at the royal chapel of St Peter ad Vincula in the Tower of London in the tradition of the day for workers at the Mint. His memorial tablet in St Gwendoline's church, Talgarth, placed there many years later, merely hints at his contribution to public life (see Fig. 22).

After Joseph's death his daughter Anna Maria stayed for a while with relatives but also made use of Place House in Lewisham before returning to the family home in the Tower of London, as she was allowed to do as a member of the Royal Mint entourage, remaining there for three years until she came of age at 21. Anna wrote to Howell in March 1766 of living a secluded life at the Tower despite visiting her uncle Thomas on several occasions. Her letters also reveal that she was quite a spirited lady with an eye on the family finances as she reported that her tenants (the Prossers) were a year and a half behind on their rent at Tredustan. In her eyes, she felt that the Prossers were obviously taking advantage of her situation after Joseph's death. Anna also mentions rents due from Penlanyfal farm near Llangorse, so it would appear that Joseph had probably invested in several properties in the area to maintain an income that sustained him throughout his various illnesses and with an eye to the future financial provision for his daughter.

There seems little doubt Anna had been well educated at home by Joseph. She records in her letters of reading Paul de Rapin's *History of England* and discoursing with her uncles on the bravery of the ancient Britons in resisting the Roman invasion and later on recounting her personal disgust at the activities of Mary Tudor. Howell, of course, continued to cajole his niece into joining his Methodist movement but she countered with her opinion that 'I hope that in our days, people have charity enough to believe though they meet with them who differ in their opinions concerning their mode of worship, yet if their notions are attended with sincerity of heart, they are equally acceptable to the divine being'. No doubt Anna Maria had been a keen observer of the discourses between her father and Howell and had learned well to be a balanced and informed individual.[8]

Joseph Harris: Scientist, Artisan, Assay Master

Sacred to the Memory of Joseph Harris, who died in the Tower of London 26th September 1764 where his remains are deposited. His great abilities and unshaken integrity were uniformly directed to the good of his country, by indefatigable attention having gained the greatest proficiency in every branch of scientific knowledge.

As an author published several tracts on different subjects, invented many instruments, monuments of his mathematical genius, yet superior to the love of fame, forbore even his name to be engraven on them. His political talents were well known the ministers of power in his days, who failed not to improve on all the wise and learned ideas which greatness of mind, candour with love of his country led him to communicate.

His reward. — In HEAVEN!

FIGURE 22 Joseph's memorial tablet in St Gwendoline's church (photograph by author)

On ornate memorial tablet in slate background with a white alabaster vase surmounting a shield with Joseph's achievements highlighted against the alabaster in black lettering. Above the shield and just below the vase is an image of a quadrant and a pair of mathematical compasses.

Eventually returning to Talgarth, she married Samuel Hughes of Tregunter. Hughes was a descendent of Peter Gunter who had been given his land by the Norman knight Bernard de Neufmarché as far back as 1083. Hughes became the Sheriff of Brecon in 1790 whilst Anna Maria inherited the Tredustan estates of her grandfather Thomas Jones through her mother. In marrying Hughes, together they united a large portion of privately owned land in the Breconshire area. When Joseph's younger brother Thomas died, he left his Tregunter estates to Anna Maria. She had her first child at the late age of thirty-five or thirty-six; in June 1784 they baptised a daughter they named Amelia Sophia. However, the estates were inherited by their second daughter (Joseph's granddaughter), Eliza Anne as it appears that Amelia died in 1790. Their third child, named Anna Maria after her mother, never married.

Eliza Anne went on to grow up and marry the wealthy landowner Roderick Gwynne of Buckland and had a daughter from this marriage, Anna Maria Eleanor Gwynne (Joseph's great-granddaughter) who went on to marry James Price Holford of Carmarthen on 4 September 1830. Holford became the Sheriff of Breconshire in 1840. Together they had a son, James Price William Gwynne-Holford (Joseph's great-great grandson) whose daughter, Louisa Mary Ermine Eleanora Gwynne Holford (Joseph's great-great-great granddaughter), married the chief constable of Monmouth, Edmund Phillip Herbert of Llansantffraed Court Abergavenny in 1865. He was a descendant of George Plantagenet, Duke of Clarence and brother of King Edward IV.[9]

Eleanora Gwynne Holford died young and is buried, along with her six-year-old son, in the church of St Bridget's at Llansantffraed. Her other son, Edmund Arthur Herbert (Joseph's great-great-great-great grandson) went to Swaziland and fought in the Boer war as commander of the 6th Inniskilling Dragoons. One of his company commanders was Lawrence Oates who would later achieve a form of immortality, dying in 1912 with Captain Robert Falcon Scott in their ill-fated Antarctic expedition. In 1898 Herbert married Ethel Rogers of Hadlow Castle, Kent and had two daughters, Eleanora and Mary Katherine Herbert (Joseph's great-great-great-great-great granddaughters).

Joseph and the Princess of Wales connection

Returning to Joseph's granddaughter Eliza Anne, she outlived Roderick Gwynne and after a suitable mourning period she became the second wife of William Alexander Maddocks MP in 1818. Eliza was just twenty-one and had returned to her Tregunter estates after Gwynne's death. Maddocks was the MP for Chippenham but owned the Tan yr Allt and Dolmelynllyn estates in north Wales where his love for the wild country and its waterfalls drew him in increasing frequency. After his marriage to Eliza, Maddocks along with Eliza's daughter Eleanora moved to Tremadog where his expertise in engineering and contracting the best surveyors led to him becoming responsible for building rail links across mid Wales and principally constructing the towns of Tremadog and Porthmadog in order to transport slate from Blaenau Ffestiniog.

Maddocks was also instrumental in draining the Traeth Mawr marshes and building the imposing stone causeway known as the Cobb across the Glaslyn estuary. However, his role as MP and his involvement in so many projects plus the rigours of fatherhood after the birth of their daughter Eliza Anne Ermine took a toll on his health. In 1826 Eliza decided that he needed to get away from it all and they went on a tour of Italy. Returning home through France, Maddocks fell ill and died in Paris on 15 September 1828 and was buried in Pere Lachaise cemetery. Eliza, now a widow for the second time at the young age of thirty-one, returned home to Brecon with both Anna Maria Eleanor Gwynne and Eliza Anne Ermine Maddocks. Eventually, the Tregunter estates were held in trust until Eliza Anne came of age. Anna Maria had of course married into the Holford family and their estates, so it seems fair that Tregunter was passed to the daughter of William Maddocks.

Eliza Anne matured and went on to marry the landowner John Webb Roche of Rochemont, Ireland. John Webb Roche was the first cousin of Edmund Roche, 1st Baron of Fermoy in Ireland. Edmund is the great-great grandfather of Diana Spencer, who upon her marriage to Charles Windsor became HRH the Princess of Wales. Joseph's family connections through his descendants are intriguing. A monarchist through and through, if the evidence of his letters on news of King George II and his family are to be believed, he would have been amused and flattered to have even a distant relationship to William, the future king of Britain.

John Webb Roche and Eliza's sons owned land at Trabolgan in Ireland, and it appears from existing archive records that they disputed land at Llysdinam, Newbridge-on-Wye in mid Wales. It seems from census records that they had moved back to Eliza's inherited estates at Tregunter, though these estates contained many outlying agricultural areas across mid and south Wales. John Webb Roche made application for the purchase of the Buckland estate of Anna Maria Eleanor Gwynne-Holford to go to his son Francis William Alexander in 1862. Francis later owned Tregunter Park and married Ellen Beatrice de Winton, third daughter of Henry de Winton, the archdeacon of Brecon. Upon his death, Francis was buried at St Gwendoline's church in Talgarth and the east window has been built in his honour.

Neither he nor his brother John Hughes Roche had any children. Leaving Buckland House, Anna Maria Eleanor Gwynne-Holford married Henry Houlton Palmer Vivian of Tregavethan, Cornwall; a family that had aristocratic connections and lands in Wales through Lord Vivian and Lord Swansea. Henry Vivian was an active member of the Royal Zoological Society and he and Anna Maria moved to his old family residence in Tregavethan, Cornwall after their marriage, which took place at Knuston Hall near Wellingborough, Northants. They had no children and Anna Maria died in December 1881.

Joseph's descendants to modern day

Joseph's line of descendants possibly continues through Eleanora and Mary Katherine Herbert, and further research on this line reveals some intriguing details. Both were born at Chichester Park, Belfast and their births registered in Ireland where their father was serving in the Inniskilling Dragoons. In 1904 he was awarded Membership of the Royal Victorian Order (MVO) for service to the Monarch, possibly connected to his leadership and defence of British interests in Ireland and South Africa during the Boer war.

Before the outbreak of the First World War, the Dragoons were stationed in India, then left for France in 1915. After the First World War, Army records reveal that Edmund was living briefly in Florence before returning to the UK. His correspondence and pay claims were sent to Moynes Court at Chepstow where he died on 29 May 1946 with the rank of retired Brigadier General. Census records show that Edmund and his daughters had been living at Moynes Court since 1910. Their old home at Llansantffraed court had been sold to Louis Rowlatt Kettle, the son of Louis Coke Kettle the Lieutenant Colonel of the 5th Inniskilling Dragoons in 1937. Whilst online genealogical records confirm the death of Eleanora at Warminster on 9 October 1983 and very little else, her sister Mary had a very chequered career.

Born in Ireland in 1903, Mary (or Maureen as she preferred) went to the Slade School of Art before enrolling at London University and gaining a degree in fine art. She did postgraduate studies at the University of Cairo before joining the civil service, working in the passport department for the British Embassy in Warsaw until the outbreak of the Second World War, whereupon she joined the air ministry in London as a translator. In September 1941 she joined the WAAF as an intelligence clerk and then went on to join the British Special Operations Executive (SOE) in March 1942. She trained alongside the famous agents Odette Sansom, Jacqueline Nearne and Lise de Baissac – all four were sent to France to aid the French resistance against the Nazi occupation.[10]

Mary was a courier in Bordeaux working for the network codenamed *scientist circuit*. It was whilst working between Paris and Bordeaux that she met her future husband Claude, the brother of fellow agent Lise de Baissac and they had a daughter they named Claudine in 1943. In February 1944 she was arrested and interned by the Gestapo, enduring harsh conditions until her release in late March 1944. She was reunited with her child and Claude and Mary married at Corpus Christi Church, London in November that year but the marriage was doomed and they did not live together long. They were finally divorced in 1959. After the war, Mary did some Arabic translation work and private tutoring in French. Suffering from depression for many years, Mary committed suicide in her garden at Uckfield in Sussex by hanging herself from the bough of an apple tree on 23 January 1983.

Her daughter Claudine is possibly still alive; she married Philip R. Pappe at Chelsea in December 1968 but it is not known if she had any children and it appears from online records that Philip may later have married Joan Stoddard and she, along with Philip, son Christopher and daughters Alison and Michelle, now live in Sonoma and San Francisco, California. Online records place Claudine in Sonoma with Philip but the date they divorced and where they lived is not known. Claudine appears to be registered as a member of a senior citizens retirement group and I am currently trying to trace this line.

The appendix (Genealogy of Joseph Harris) following this chapter traces the lineage.

It would be a great pity if Joseph's distinguished line has come to an end. Nevertheless, Joseph would have been proud of the civic traditions that his successors and their families contributed to. He may have been amused and gratified that his descendants returned to the Brecon area and played a large part in the social, political and cultural activities of their day.

Appendix A

GENEALOGY OF JOSEPH HARRIS

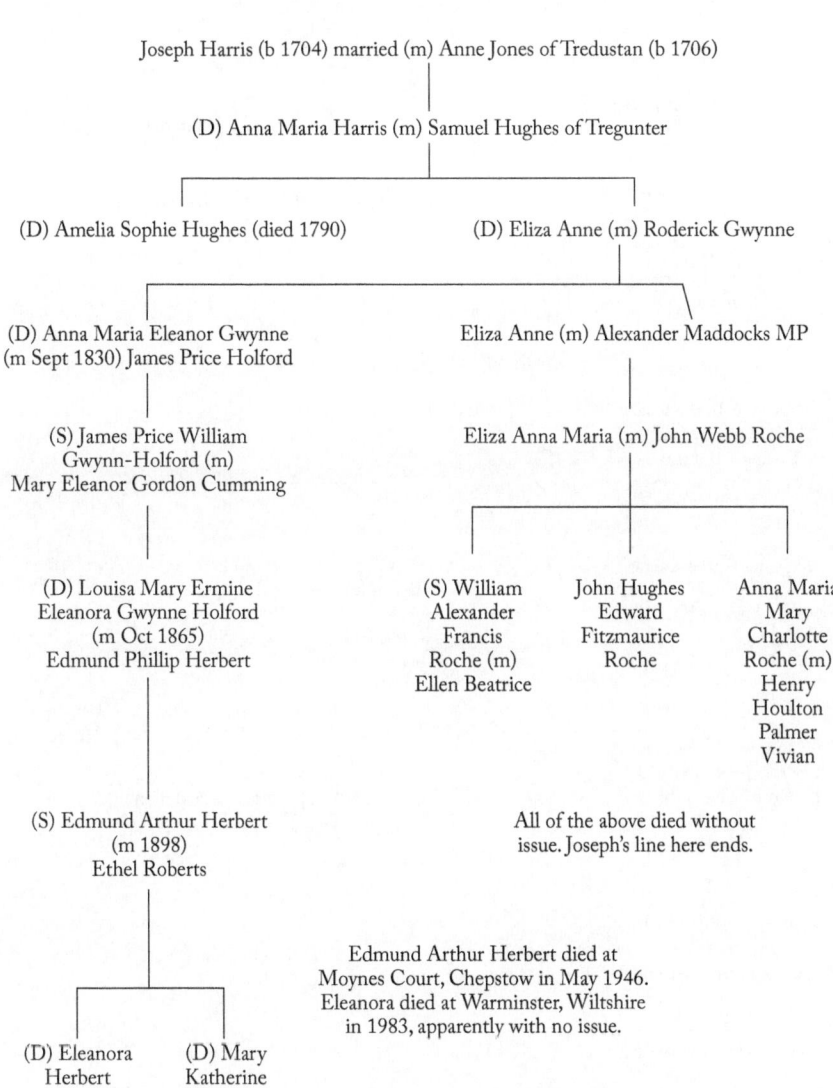

SCIENTISTS OF WALES

Joseph Harris and the Diana Spencer connection

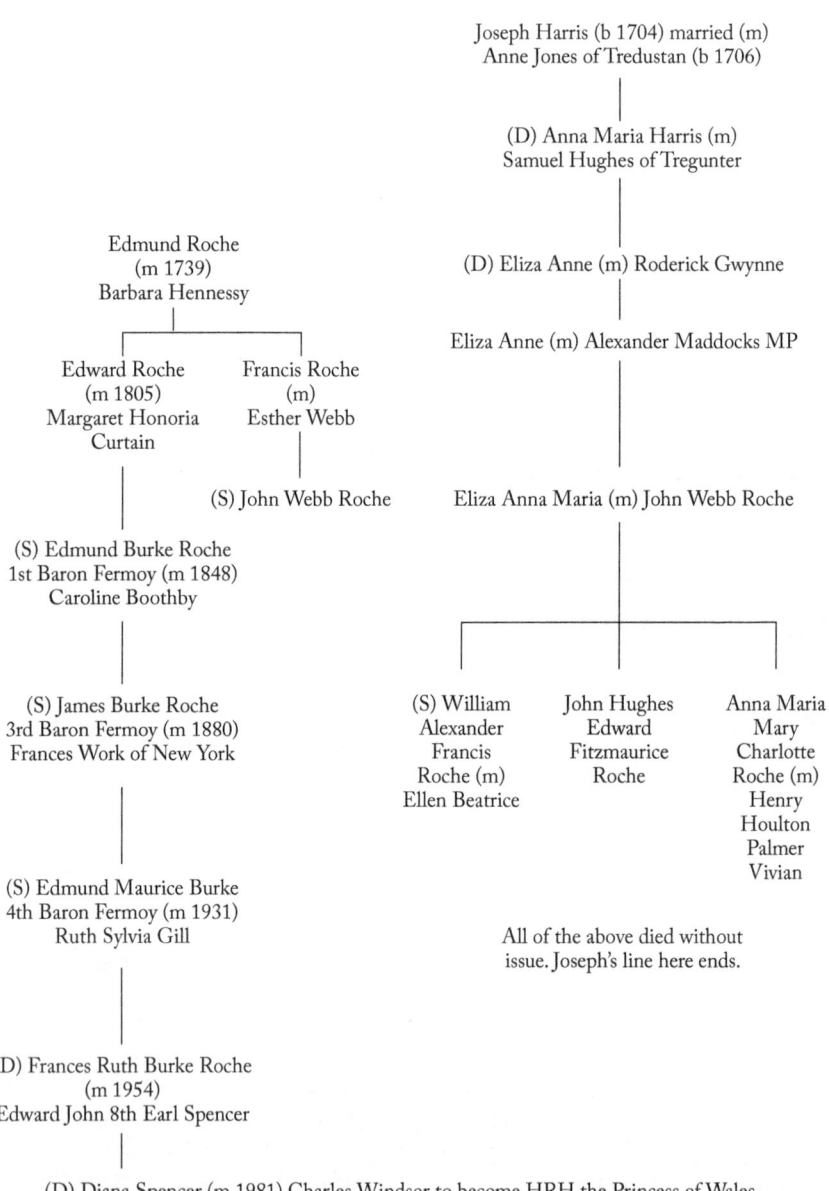

148

Appendix B

TIMELINE OF JOSEPH HARRIS

1704   Joseph Harris born in Trefecca to Howell and Susannah Powell (they would change their name to Harris)

1715   Apprenticed as a blacksmith to his uncle Thomas Powell

1725   Moves to London with letters of introduction to John Senex and Edmond Halley. Starts apprenticeship to Senex

1725   Undertakes his first Caribbean voyage checking stellar positions

1727   Returns to London and prepares chart of the Southern Stars

1728   His paper on his Caribbean observations published by Halley in *Philosophical Transactions*

1730   Publishes *The Description and Use of the Globes, and the Orrery*

1730   Prepares materials for the 1731 publication of *A Treatise of Navigation*

1730   Undertakes his second Caribbean voyage

1731   Returns to London and publishes *A Treatise of Navigation*

1732   Stays with John Conduitt MP

1733   Becomes mathematics teacher to son of John Knight MP

1733   Occasional demonstrator at the Royal Society

1735	Returns to Trefecca to urge Howell Harris to matriculate at Oxford
1736	Appointed deputy Assay Master at the Royal Mint, London under Hopton Haynes
1736	Marries Anne Jones at St Benet's Welsh Church in London
1739	Demonstrates the uses of globes at the Royal Society
1741	Joseph assays the metals for the Copley Medal of the Royal Society
1746	New silver coins in circulation assayed by Joseph and designed by Sigismund Tanner
1748	Joseph and Anne move into the Tower of London
1749	Appointed Assay Master of the Mint
1753	Joseph awarded a royal pension by King George II
1757	New recoinage overseen by Joseph Harris
1757	Stanesby Alchorne becomes Joseph's deputy Assay Master
1758	Joseph publishes parts I and II of *Essay on Money and Coins*
1758	Joseph becomes a consultant to the Carysfort Commission
1761	Watches a total lunar eclipse and the transit of Venus from Trefecca
1763	Death of Joseph's wife Anne
1764	Joseph dies in September and is buried in the church of St Peter ad Vincula in the Tower of London
1768	Stanesby Alchorne produces Joseph's *Essay on Money and Coins* part III based on his notes and observations
1775	Joseph's last work *A Treatise of Optics* is published

NOTES

Chapter 1

1. J. E. Lloyd and R. T. Jenkins, *The Dictionary of Welsh Biography* (Cardiff: University of Wales Press, 1992).
2. P. Thomas, *The Remaking of Wales in the Eighteenth Century* (Maidenhead: Open University Press, 2010), pp. 3–5.
3. Thomas, *The Remaking of Wales in the Eighteenth Century*, pp. 3–5.
4. Thomas, *The Remaking of Wales in the Eighteenth Century*, pp. 3–5.
5. Thomas, *The Remaking of Wales in the Eighteenth Century*, pp. 3–5.
6. Thomas, *The Remaking of Wales in the Eighteenth Century*, pp. 3–5.
7. Thomas, *The Remaking of Wales in the Eighteenth Century*, pp. 3–5.
8. Lloyd and Jenkins, *The Dictionary of Welsh Biography*.
9. M. H. Jones, 'Joseph Harris: An Assay Master of the Mint', *Journal of the Welsh Bibliographical Society*, 3 (6) (July 1929), 256–67.
10. R. Bennett, *The Early Life of Howell Harris* (Edinburgh: Banner of Truth Trust, 1962), p. 4.
11. *https://www.historyofparliamentonline.org/volume/1715-1754/member/jones-roger-1691-1741*.

Chapter 2

1. Emrys Jones, *The Welsh in London 1500–2000* (Cardiff: University of Wales Press, 2001), pp. 55–7.
2. B. S. Schlenther and E. M. White, *Calendar of the Trevecka Letters* (Aberystwyth: National Library of Wales, 2003), p. 1.
3. M. H. Jones, 'Joseph Harris: An Assay Master of the Mint', *Journal of the Welsh Bibliographical Society*, 3/6 (July 1929), 256–67.
4. Jones, 'Joseph Harris', 256–67.
5. M. Rees, *The Life of John Flamsteed Britain's First Astronomer* (Amber Valley: Royal Amber Valley Press, 2017), p. 75.
6. Wolfgang Steinicke, 'William Herschel, Flamsteed Numbers and Harris's Star Maps', *Journal for the History of Astronomy*, 45 (July 2014).

7. E. Brooke-Hitching, *The Sky Atlas* (New York: Simon & Schuster Press, 2019), p. 86.
8. Jones, 'Joseph Harris', 256–67.

Chapter 3

1. E. Halley, 'Astronomical Observations at Vera Cruz by Mr Joseph Harris', *Philosophical Transactions of the Royal Society* (hereafter *Phil. Trans.*), 35 (1735), 388–9.
2. B. S. Schlenther and E. M. White, *Calendar of the Trevecka Letters* (Aberystwyth: National Library of Wales, 2003), p. 4.
3. Vera Lee Brown, 'The South Sea Company and the Contraband Trade', *The American Historical Review*, 31/4 (July 1926), 662–78.
4. Schlenther and White, *Calendar of the Trevecka Letters*, p. 6.
5. Halley, 'Astronomical Observations at Vera Cruz', 388–9.
6. Halley, 'Astronomical Observations at Vera Cruz', 388–9.
7. Halley, 'Astronomical Observations at Vera Cruz', 388–9.
8. Halley, 'Astronomical Observations at Vera Cruz', 388–9.
9. Dava Sobel, *Longitude* (London: HarperCollins, 2005), p. 17.
10. D. J. Bryden, 'The Jamaican Observatories of Colin Campbell, F.R.S. and Alexander Macfarlane, F.R.S.', *Notes and Records of the Royal Society of London*, 24 (2 April 1970), 261–72.
11. Schlenther and White, *Calendar of the Trevecka Letters*, p. 7.
12. Schlenther and White, *Calendar of the Trevecka Letters*, p. 9.
13. Bryden, 'The Jamaican Observatories of Colin Campbell, F.R.S. and Alexander Macfarlane, F.R.S.', 261–72.
14. C. Amory-Mazaudier, 'Electrical Currents in the Earth's Environment – Some Historical Aspects', IAGA meeting in Buenos Aires, published in *GEOACTA*, 21 (1994), 1–15.
15. J. Harris and G. Graham, 'An Account of some Magnetical Observations Made in the Months of May, June and July 1732 in the Atlantick or Western Ocean; Also the Description of a Waterspout by Mr Joseph Harris, communicated by Mr George Grahame FRS', *Phil. Trans.* (1733), 38, 75–9.
16. Bryden, 'The Jamaican Observatories of Colin Campbell, F.R.S. and Alexander Macfarlane, F.R.S.', 261–72.
17. Bryden, 'The Jamaican Observatories of Colin Campbell, F.R.S. and Alexander Macfarlane, F.R.S.', 261–72.
18. Harris and Graham, 'An Account of some Magnetical Observations', 75–9.
19. Schlenther and White, *Calendar of the Trevecka Letters*, p. 14.
20. *Derby Mercury*, 26 November 1756, p. 2.

Chapter 4

1. M. H. Jones, 'Joseph Harris: An Assay Master of the Mint', *Journal of the Welsh Bibliographical Society*, 3/6 (July 1929), 256–67.

2. D. J. Warner, *The Sky Explored: Celestial Cartography 1500–1800* (Kansas, 1979) p. 134.
3. Warner, *The Sky Explored*, p. 134.
4. Wolfgang Steinicke, 'William Herschel, Flamsteed Numbers and Harris's Star Maps', *Journal for the History of Astronomy*, 45 (July 2014).
5. Joseph Harris, *A Treatise of Navigation* (London, 1730), Introduction.
6. Harris, *A Treatise of Navigation*, pp. 1–4.
7. Harris, *A Treatise of Navigation*, pp. 5–7.
8. M. Kerr, *The Navy in My Time* (London: Rich & Cowan, 1933), pp. 17–18.
9. Kerr, *The Navy in My Time*, pp. 17–18.
10. Jones, 'Joseph Harris: An Assay Master of the Mint', 256–67.
11. John Harris, *The Description and Uses of the Celestial and Terrestrial Globes; and of Collins's Pocket-Quadrant* (London, 1703).
12. Joseph Harris, *The Description and Use of the Globes, and the Orrery* (London: G. Hawkins, 1731).
13. J. J. O'Connor and E. F. Robertson, 'Thomas Digges, a Brief Biography' (St Andrews: University of St Andrews, 2002).
14. Nick Kanes, *Star Maps: History, Artistry, and Cartography* (Chichester: Springer Praxis Books, 2007).
15. Harris, *The Description and Use of the Globes, and the Orrery*, pp. 2–10.
16. Harris, *The Description and Use of the Globes, and the Orrery*, pp. 2–10.
17. Harris, *The Description and Use of the Globes, and the Orrery*, pp. 11–16.
18. Harris, *The Description and Use of the Globes, and the Orrery*, pp. 11–16.
19. Harris, *The Description and Use of the Globes, and the Orrery*, pp. 25–30.
20. Harris, *The Description and Use of the Globes, and the Orrery*, pp. 25–30.
21. Jones, 'Joseph Harris: An Assay Master of the Mint', 256–67.
22. Jones, 'Joseph Harris: An Assay Master of the Mint', 256–67.
23. Jones, 'Joseph Harris: An Assay Master of the Mint', 256–67.
24. Jones, 'Joseph Harris: An Assay Master of the Mint', 256–67.
25. J. Harris, 'An Account of an Improvement on the Terrestrial Globe', *Phil. Trans.*, 41 (1739), 321–6.
26. Harris, *The Description and Use of the Globes, and the Orrery*, p. 172.

Chapter 5

1. J. Craig, 'The Royal Society and the Royal Mint', *Records of the Royal Society of London*, 19 (2 December 1964), 156–67.
2. Craig, 'The Royal Society and the Royal Mint', 156–67.
3. Craig, 'The Royal Society and the Royal Mint', 156–67.
4. Alex Bellotti, 'Lions and monkeys and bears used to live in the Tower of London', *https://londonist.com/london/history/the-tower-of-london-menagerie*
5. A. Rupert Hall, *Isaac Newton: Adventurer in Thought* (Cambridge: Cambridge University Press, 1993), p. 337.
6. Brodie Waddell, 'The Economic Crisis of the 1690s in England', *The Historical Journal*, 66/2 (March 2023).

7. Craig, 'The Royal Society and the Royal Mint', 156–67.
8. P. Mathias, *The Transformation of England: Essays in the Economic and Social Transformation of England in the Eighteenth Century* (London: London University Paperbacks, 1979), p. 194.
9. P. Wakelin and J. Day, 'Joseph Harris in Bristol, 1748', *BIAS Journal*, 15 (1982).
10. J. Craig, 'The Royal Society and the Royal Mint'.
11. Joseph Harris, *An Essay upon Money and Coins* (London, 1758), p. 7.
12. Mathias, *The Transformation of England*, p. 194.
13. M. Lessen, 'Harris, Alchorne and "An Essay"', *British Numismatic Society Journal*, 62 (1992).
14. Joseph Harris, *Letter to Martin Folkes at the Royal Society*, British Library MS ADD 4441f.80 1741.

Chapter 6

1. *The Trevecca letters in 3 volumes* held at the National Library of Wales, Aberystwyth, personal research.
2. B. S. Schlenther and E. M. White, *Calendar of the Trevecka Letters* (Aberystwyth: National Library of Wales, 2003).
3. Schlenther and White, *Calendar of the Trevecka Letters*.
4. T. Beynon, *Howell Harris's Visits to London* (Aberystwyth: Cambrian News Press, 1960).
5. Schlenther and White, *Calendar of the Trevecka Letters*.
6. R. Bennett, *Howell Harris and the Dawn of Revival* (Bridgend: Evangelical Press of Wales, 1987).
7. Schlenther and White, *Calendar of the Trevecka Letters*.
8. Schlenther and White, *Calendar of the Trevecka Letters*.
9. *The Trevecca letters in 3 volumes*, National Library of Wales, personal research.
10. Schlenther and White, *Calendar of the Trevecka Letters*.
11. D. E. Pike, *https://daibach-welldigger.blogspot.com/2018/09/howel-harris-trefeca-welsh-utopia.html*.
12. F. Cook, *Selina, Countess of Huntingdon* (Edinburgh: Banner of Truth Trust, 2002).
13. Schlenther and White, *Calendar of the Trevecka Letters*.
14. Brecknockshire Agricultural Society, *https://www.breconcountyshow.co.uk/aboutus.html*
15. G. Davies, 'Trevecka (1706–1964)', *Brycheiniog*, 15 (1971), 99.
16. Schlenther and White, *Calendar of the Trevecka Letters*.
17. Schlenther and White, *Calendar of the Trevecka Letters*.
18. D. E. Pike, *https://daibach-welldigger.blogspot.com/2018/09/howel-harris-trefeca-welsh-utopia.html*
19. *The Trevecca letters in 3 volumes* held at the National Library of Wales, personal research.
20. David Ceri Jones, *A Glorious Work in the World: Welsh Methodism and the International Evangelical Revival 1735–1750* (Cardiff: University of Wales Press, 2004).
21. Schlenther and White, *Calendar of the Trevecka Letters*.

22. M. J. Levy, *Perdita, the Memoirs of Mary Robinson* (London: Peter Owen Publishers Ltd, 1994).
23. Levy, *Perdita*.

Chapter 7

1. John Proby, *https://www.historyofparliamentonline.org/volume/1754-1790/member/proby-john-1720-72*.
2. John Proby, *The Carysfort Commission and the Wine Gallon*, Report from the Committees of the House of Commons (1737–65), Vol. II.
3. William Waterston, *A Cyclopaedia of Commerce, Mercantile Law and Finance* (Henry Bohn: London, 1846), p. 64.
4. Carysfort, *Report from the Committee appointed to enquire into the standards of weights and measures* (House of Commons, 1758), p. 44.
5. Carysfort, *Report from the Committee appointed to enquire into the standards of weights and measures* (House of Commons, 1758), p. 51.
6. Carysfort, *Report from the Committee appointed to enquire into the standards of weights and measures* (House of Commons, 1758), p. 59.
7. Carysfort, *Report from the Committee appointed to enquire into the standards of weights and measures* (House of Commons, 1758), p. 44.
8. John Proby, *https://www.historyofparliamentonline.org/volume/1754-1790/member/proby-john-1720-72*.

Chapter 8

1. Colin Ronan, *Edmond Halley: Genius in Eclipse* (London: Macdonald & Co., 1970), pp. 113–17.
2. Ronan, *Edmond Halley*, pp. 113–17.
3. Michael Wright, 'The Ordeal of Guillaume Le Gentil', *Sidereal Times*, Princeton, February 2012.
4. Joseph Harris, 'Letter from Joseph Harris to his brother Howell Harris', National Library of Wales Archives (1761).
5. B. S. Schlenther and E. M. White, *Calendar of the Trevecka Letters* (Aberystwyth: National Library of Wales, 2003).
6. Schlenther and White, *Calendar of the Trevecka Letters*.
7. *Brycheiniog*, Vol. 41 (2005).
8. J. Stanesby-Moody, *Brycheiniog*, 41 (2005).

Chapter 9

1. William Waterston, *A Cyclopaedia of Commerce, Mercantile Law and Finance* (Henry Bohn: London, 1846).
2. N. Mayhew, 'Prices in England 1170–1750', *Past and Present*, 219 (2013).
3. M. Allen, *Mints and Money in Medieval England* (Cambridge: Cambridge University Press, 2012).
4. Allen, *Mints and Money in Medieval England*.
5. Mayhew, 'Prices in England 1170–1750'.

6. J. Conduitt, *Treatise on Gold and Silver Standards* (London, 1774).
7. Joseph Harris, *An Essay upon Money and Coins* (London, 1757), frontispiece.
8. Harris, *Essay upon Money and Coins*, p. 53.
9. Harris, *Essay upon Money and Coins*, p. 98.
10. Adam Smith, *An Inquiry into the Nature and Causes of the Wealth of Nations* (Oxford: Oxford University Press, 2008), p. 15.
11. M. Lessen, 'Harris, Alchorne and "An Essay"', *The British Numismatic Journal*, 62 (1992).

Chapter 10

1. Joseph, Harris, *A Treatise of Optics* (G. Hawkins: London, 1775), Introduction.
2. Harris, *A Treatise of Optics*, p. 143.
3. Harris, *A Treatise of Optics*, p. 30.
4. Harris, *A Treatise of Optics*, p. 45.
5. Harris, *A Treatise of Optics*, p. 121.
6. Harris, *A Treatise of Optics*, p. 88.
7. Harris, *A Treatise of Optics*, p. 129.
8. Roy Porter, *The Greatest Benefit to Mankind: A Medical History of Humanity from Antiquity to the Present* (London: HarperCollins, 1997), p. 275.
9. Harris, *A Treatise of Optics*, p. 334.
10. Harris, *A Treatise of Optics*, p. 183.
11. Harris, *A Treatise of Optics*, p. 391.
12. Harris, *A Treatise of Optics*, p. 319.

Chapter 11

1. Howell Harris, *Personal diaries*, National Library of Wales, Aberystwyth.
2. Joseph Harris, *Personal letters*, National Library of Wales, Aberystwyth.
3. J. E. Lloyd and R. T. Jenkins, *The Dictionary of Welsh Biography* (Cardiff: University of Wales Press, reprint, 1992).
4. G. Davies, 'Trevecka (1706–1964)', *Brycheiniog*, 15 (1971), 99.
5. Emrys Jones, *The Welsh in London 1500–2000* (Cardiff: University of Wales Press, 2001), pp. 21–2.
6. H. Owen, 'Additional Letters of the Morrises of Anglesey', *Honourable Society of Cymmrodorion*, 2 (1947–9), 981.
7. Davies, 'Trevecka (1706–1964)', 99.
8. Joseph Harris, *Personal letters*, National Library of Wales.
9. Marquis of Ruvigny, *The Plantagenet Roll of the Blood Royal* (Heritage Classics: London, 1911).
10. Michael Richard Daniell Foot and Jean-Louis Crémieux-Brilhac, *Of the English in Resistance: British Secret Service of Action (SOE) in France 1940–1944* (Tallandier, 2008).

BIBLIOGRAPHY

Allen, M., *Mints and Money in Medieval England* (Cambridge: Cambridge University Press, 2012).

Amory-Mazaudier, C., 'Electrical Currents in the Earth's Environment – Some Historical Aspects', IAGA meeting in Buenos Aires, published in *GEOACTA*, 21 (1994), 1–15.

Bellotti, Alex, 'The animals of the Tower menagerie', https://londonist.com/london/history/the-tower-of-london-menagerie.

Bennett, R., *The Early Life of Howell Harris* (Edinburgh: Banner of Truth Trust, 1962).

Bennett, R., *Howell Harris and the Dawn of Revival* (Bridgend: Evangelical Press of Wales, 1987).

Beynon, T., *Howell Harris's Visits to London* (Aberystwyth: Cambrian News Press, 1960).

Boydell and Brewer, *Parliamentary History Online*, https://www.historyofparliamentonline.org/volume/1715-1754/member/jones-roger-1691-1741

Brooke-Hitching, E., *The Sky Atlas* (New York: Simon & Schuster Press, 2019).

Brown, Vera Lee, 'The South Sea Company and Contraband Trade', *The American Historical Review*, 31/4 (July 1926), 662–78.

Bryden, D. J., 'The Jamaican Observatories of Colin Campbell, F.R.S. and Alexander Macfarlane, F.R.S.', *Notes and Records of the Royal Society of London*, 24 (2 April 1970).

Carysfort, *Report from the Committee appointed to enquire into the standards of weights and measures* (House of Commons, 1758).

Conduitt, J., *Treatise on Gold and Silver Standards* (London, 1774).

Cook, F., *Selina, Countess of Huntingdon* (Edinburgh: Banner of Truth Trust, 2002).

Craig, J., 'The Royal Society and the Royal Mint', *Records of the Royal Society of London*, 19 (1 December 1964).

Davies, G., 'Trevecka (1706–1964)', *Brycheiniog*, 15 (1971), 99.

Foot, Michael Richard Daniell and Crémieux-Brilhac, Jean-Louis, *Of the English in Resistance: British Secret Service of Action (SOE) in France 1940–1944* (Tallandier, 2008).

Foucault, Didier, 'The invention of the solar system (16th–18th centuries), a recent astronomical concept: the planetary system. Epistemological relevance and methodological precautions', *The Past and Present of Aeronautics and Space*, 4 (2018).

Hall, A. Rupert, *Isaac Newton, Adventurer in Thought* (Cambridge: Cambridge University Press, 1993).

Halley, E., 'A Description of the Passage of the Shadow of the Moon over England as it was Observed in the late Total Eclipse of the Sun April 22d, 1715', Broadsheet by Senex (1715).

Halley, E., 'Astronomical Observations at Vera Cruz by Mr Joseph Harris', *Philosophical Transactions of the Royal Society* (hereafter *Phil. Trans.*) (1735).

Harris, Howell, *Personal diaries*, National Library of Wales, Aberystwyth.

Harris, John, *The Description and Uses of the Celestial and Terrestrial Globes; and of Collins's Pocket-Quadrant* (London, 1703)

Harris, Joseph, *Personal letters*, National Library of Wales, Aberystwyth.

Harris, Joseph, *A Treatise of Navigation* (London, 1730).

Harris, Joseph, *The Description and Use of the Globes, and the Orrery* (London: G. Hawkins, 1731).

Harris, Joseph, *An Essay upon Money and Coins* (London, 1757/8).

Harris, Joseph, *A Treatise of Optics* (G. Hawkins: London, 1775).

Harris, J. and Graham, G., 'An Account of some Magnetical Observations Made in the Months of May, June and July 1732 in the Atlantick or Western Ocean; Also the Description of a Waterspout by Mr Joseph Harris, communicated by Mr George Grahame FRS', *Phil. Trans.* (1733).

Headrick, Daniel R., *When Information Came of Age: Technologies of Knowledge in the Age of Reason and Revolution, 1700–1850* (Oxford: Oxford University Press, 2000).

Jones, David Ceri, *A Glorious Work in the World: Welsh Methodism and the International Evangelical Revival 1735–1750* (Cardiff: University of Wales Press, 2004).

Jones, Emrys, *The Welsh in London 1500–2000* (Cardiff: University of Wales Press, 2001).

Jones, M. H., 'Joseph Harris: An Assay Master of the Mint', *Journal of the Welsh Bibliographical Society* (July 1929).

Kanes, Nick, *Star Maps: History, Artistry, and Cartography* (Chichester: Springer Praxis Books, 2007).

Kerr, M., *The Navy in My Time* (London: Rich & Cowan, 1933).

Lessen, Marvin, 'Harris, Alchorne and "An Essay"', *The British Numismatic Journal*, 62 (1992).

Levy, M. J., *Perdita, the Memoirs of Mary Robinson* (London: Peter Owen Publishers Ltd, 1994).

Lloyd, J. E. and Jenkins, R. T., *Dictionary of Welsh Biography* (Cardiff: University of Wales Press, reprint, 1992).

Mathias, P., *The Transformation of England: Essays in the Economic and Social Transformation of England in the Eighteenth Century* (London: London University Paperbacks, 1979).

Mayhew, N., 'Prices in England 1170–1750', *Past and Present*, 219 (2013).

Moody, J., *Brycheiniog*, 41 (2005).

O'Connor, J. J. and Robertson, E. F., 'Thomas Digges, a Brief Biography' (University of St Andrews, 2002), available at *https://mathshistory.st-andrews.ac.uk/Biographies/Digges/*

Owen, H., 'Additional Letters of the Morrises of Anglesey', *Honourable Society of Cymmrodorion*, 2 (1947–9), 981.

Pike, D. E., *https://daibach-welldigger.blogspot.com/2018/09/howel-harris-trefeca-welsh-utopia.html*

Porter, Roy, *The Greatest Benefit to Mankind: A Medical History of Humanity from Antiquity to the Present* (London: HarperCollins, 1997).

Proby, J., *The Carysfort Commission and the Wine Gallon*, Report from the Committees of the House of Commons (1737–65), Vol. II.

Rees, M., *The Life of John Flamsteed, Britain's First Astronomer* (Amber Valley: Royal Amber Valley Press, 2017).

Ronan, Colin, *Edmond Halley: Genius in Eclipse* (London: Macdonald & Co., 1970).

Ruvigny, Marquis of, *The Plantaganet Roll of the Blood Royal* (Heritage Classics: London, 1911).

Schlenther, B. S. and White, E. M., *Calendar of the Trevecka Letters* (Aberystwyth: National Library of Wales, 2003).

Smith, Adam, *An Inquiry into the Nature and Causes of the Wealth of Nations* (Oxford: Oxford University Press, 2008).

Sobel, Dava, *Longitude* (London: HarperCollins, 2005).

Steinicke, Wolfgang, 'William Herschel, Harris' Star Maps and Flamsteed Numbers', *Journal for the History of Astronomy*, 45 (July 2014).

Thomas, P., *The Remaking of Wales in the 18th Century* (Maidenhead: Open University Press, 2010).

Waddell, Brodie, 'The Economic Crisis of the 1690s in England', *The Historical Journal*, 66/2 (March 2023).

Wakelin, P. and Day, J., 'Joseph Harris in Bristol, 1748', *BIAS Journal*, 15 (1982).

Warner, D. J., *The Sky Explored: Celestial Cartography 1500–1800* (Kansas, 1979).

Waterston, William, *A Cyclopaedia of Commerce, Mercantile Law and Finance* (Henry Bohn: London, 1846).

Whiston, W. and Halley, E., *18th Century eclipse maps* (London: Science Museum reprints, 2006).

Wright, Michael, 'The Ordeal of Guillaume Le Gentil', *Sidereal Times*, Princeton, 6 February 2012, available at *https://princetonastronomy.com/2012/02/06/the-ordeal-of-guillaume-le-gentil/*

INDEX

A
Aberystwyth castle 62
Alchorne, Stanesby 71
Arundel, Richard MP 63, 71, 95
Astronomical Unit (AU) 103
Atlas Britannica Coelestis 28, 48, 50
Avicenna 105

B
Bideford 88, 109
Bilson-Legge, Henry 71, 101, 124
Black River 38
Bliss, Nathaniel 106
Boodles club 125, 138
Bradley, James 38, 39, 42
Brecknockshire Agricultural Society 1, 85, 139
Brecon Market 85
Bristol 64, 69, 70, 154, 159

C
Campbell, Colin 38, 39, 42, 48, 53
camera obscura 42, 127, 128, 131
Caribbean 17, 32, 36, 37, 38, 41, 48, 50, 53, 57, 74, 76, 82, 136
Carmarthen 3, 5, 124
Carysfort (commission) 72, 97, 100, 101, 120, 150
Catalogus Stellarum Australium 28, 48
Celsius, Anders 41
Chappe d'Auteroche 107
Chester 83
Coggs, John 36, 51, 135

comets 56
compass (azimuth) 25, 29, 31, 36, 42, 51, 141
Conduitt, Catherine 63
Conduitt, John MP 63, 122, 149
Copernicus, Nicolaus 6
copper (coins) 67, 69 70, 109, 122
counterfeiting 61, 62, 64, 67
Crabtree, William 105
Cranbury Park 43, 63
Crane Court 10, 20, 24, 57, 58
Cymmrodorion Society 18, 137, 138

D
Digges, Thomas 54
dissenting schools 8, 9, 10
Dixon, Jeremiah 106
Doppelmeyr, Gabriel 55

E
eclipse (lunar) 42, 109, 113, 118, 139, 150
eclipse (solar) 11, 13, 26, 27, 35, 36, 58, 133
Essay Upon Money and Coins 66, 70, 71, 81, 101, 119, 122, 124, 125, 126
Evans, Evan 138
Evans, Hugh 10
Exeter 64

F
Flamsteed, John 27, 29, 35, 47, 48, 50, 52
Fleet Street 19, 20, 23, 24, 29, 32, 51, 58, 93, 132

Forestaff 25, 31, 51
Fontenelle, Bernard Bouvier de 56

G
George II (King) 5, 9, 72, 87
Globes and the Orrery 20, 37, 47, 53, 55, 59, 128
gold (standard) 66, 69
Goodman's Inn Fields 18
Gosfield Hall 43, 45
Graham, George 15, 21, 40, 58, 135
Great Yarmouth 88
Greenwich 16, 27, 35, 49, 50, 51, 58, 105, 106, 108, 109, 111, 112
Gregorian calendar 4, 5
Gregory, David 55

H
Hackney 66, 68
Hadley, John 58
Halley, Edmond 13, 14, 15, 20, 21. 35, 37, 39, 47, 49, 50–7 64, 109, 116–17
Harris, Howell 4, 82, 83, 85–97
Harris, Susannah 4, 82, 83
Harris, Thomas 86, 91–4
Harrison, John 15, 58
Haynes, Hopton 66, 68, 150
Heath, Thomas 20, 25, 31, 38, 51
Herschel, William 50, 136
Horrocks, Jeremiah 105
Hughes, Samuel 142
Huntingdon, Lady 89, 92, 95

J
Jamaica 31, 34, 38–42, 53, 57, 136
Jenkinson, Mary 66
Jones, Anne 32, 65, 66, 74, 84
Jones, Roger 14, 24, 26
Jones, Thomas 11, 12, 32, 53, 66
Jurin, James 130, 131

K
Kepler, Johannes 6, 56, 105
Kerr, Mark 52, 53
Knight (Anne) 43–4
Knight, John MP 43–4, 57

L
Lalande, Jérôme 86
latitude 39, 51, 58, 59, 111
Le Gentil Guillaume 106–8
Lewisham 68, 140
Lima 68, 109
Lincoln's Inn Fields 18
Lizard, Cornwall 57
Llanigon 10
Llwyn Llwyd 9, 10, 11
Locke, John 56, 71, 124
longitude 24–7, 32, 35, 37, 40, 50, 51, 57, 108, 109, 111, 112, 117
Longitude Act 37

M
Macclesfield (Earl) 25, 104, 112
Madagascar 107
Manilla 107
Mason, Charles 106
Maskelyne Nevil 106
Maupertuis Pierre 41
Mayan 105
Mayer, Tobias 131
menagerie 65, 66
Meridian (Trefecca) 109
Methodism 76, 79, 80, 86, 89, 90
Monometallism 66, 67, 71, 120
Morris, Lewis 18, 25, 137, 138
Morris, Richard 18, 25, 137, 138
Mudge, Thomas 58

N
Newcomen engine 70
Newton, Isaac 6, 15, 23, 25, 27, 40, 62, 63–6, 120, 132

Navigation (Treatise of) 38, 48, 51–3
Norwich 64, 88
Nugent, Richard 81
Nugent, Robert 45

O
Owen, Goronwy 138

P
Pelham-Holles (Newcastle duke of) 100, 106
perspective box 131
Philosophical Transactions 23, 35, 42, 43, 59, 131
Prince Frederick 32, 34, 37
Place House 68, 140
Pondicherry 106–8
Portsmouth 44, 47, 52, 53, 54, 57
Powell, Howel 3
Powell, Thomas 9
Price, David 9, 10
Price, Solomon 18
Principia 6, 25
Proby, John 95
   see also Carysfort

Q
quadrant 21, 24, 35, 36, 54, 108, 111, 141
Queen Anne 22, 24, 32

R
Reunion Island 107
Robinson, Thomas 92
Robinson, Mary (Perdita) 94–6
rolling mill 64
Royal Mint 43–7, 59, 60–72
Royal Society 13, 17, 20–5, 27, 38, 42, 43, 50, 57, 58, 63, 78, 91, 96, 101, 106, 112

S
St Benet's (church) 75
St Gwendoline's (church) 4

Senex, John 11, 13, 19, 23, 26–9
Short, James 29, 59, 106, 112
Shovell, Cloudesley 52
Shute Samuel 14
Sissons John 98, 132
Smith Adam 122, 125, 139
South Sea Bubble 33, 123
South Sea Company 32
Solar System 24, 54, 55, 57, 106, 128
Spotswood 32–6
Stanesby-Moody, Jenny 108, 112

T
Talgarth 2, 4, 6, 73, 75–80, 90, 92, 112, 136, 140
Tredustan 1, 32, 53, 65, 66, 84, 112, 140, 142
Tregunter 92, 142
Trefecca 10, 14, 42, 58, 70–94
Tompion, Thomas 135
Torrington, Viscount 51

U
Uranus 50, 57, 136
Utrecht (Treaty of) 32

V
Vera Cruz 31–3, 36, 105
Venus 42, 54, 55, 103–18

W
Wesley, John 83, 86
Whewell, William 56
Whiston William 13, 27, 54
Whitefield, George 80, 83–4, 89–91
William Williams 10, 76–8
Winthrop John 106
witchcraft 63
Wolfe, James (general) 88
Wright, Thomas 20, 51, 52, 93

Y
York 64, 94